I0043654

Roger Sherman Tracy

The Outlines of Anatomy, Physiology, and Hygiene

Roger Sherman Tracy

The Outlines of Anatomy, Physiology, and Hygiene

ISBN/EAN: 9783744670371

Printed in Europe, USA, Canada, Australia, Japan

Cover: Foto ©berggeist007 / pixelio.de

More available books at **www.hansebooks.com**

THE OUTLINES

OF

ANATOMY, PHYSIOLOGY, AND HYGIENE

BEING AN EDITION OF

THE ESSENTIALS OF ANATOMY, PHYSIOLOGY, AND HYGIENE

REVISED TO CONFORM TO THE LEGISLATION MAKING THE
EFFECTS OF ALCOHOL AND OTHER NARCOTICS UPON THE HUMAN SYSTEM
A MANDATORY STUDY IN PUBLIC SCHOOLS

BY

ROGER S. TRACY, M. D.

REGISTER OF RECORDS OF THE NEW YORK CITY HEALTH DEPARTMENT;
AUTHOR OF "HAND-BOOK OF SANITARY INFORMATION
FOR HOUSEHOLDERS"

NEW YORK ·:· CINCINNATI ·:· CHICAGO
AMERICAN BOOK COMPANY

INDORSEMENT.

The "Authorized Physiology Series" consists of :

No. I.—"Health for Little Folks." For Primary Grades.

II.—"Lessons in Hygiene,"
A special edition of "How we Live," } For Intermediate Grades.
by Johonnot and Bouton.

III.—"The Outlines of Anatomy, Physi- } For High - Schools and
ology, and Hygiene," by Roger S. } Advanced Classes in
Tracy, M. D. } Common Schools.

In this series good judgment has been shown in the selection of facts that should be taught each grade, and in representing these facts in language which the pupils of the grades for which the books are designed can comprehend.

The treatment of the subjects of anatomy and physiology in the high-school book covers the usual ground ; in the books of the lower grades it does not unduly preponderate, but it is abundantly ample to enable the pupil to comprehend the topic which is the real object of the study, viz., the laws of health, and the nature of alcoholic drinks and other narcotics, and their effects upon the human system. The treatment of this part of the subject, in each book of the series, is full, clear, and in harmony with the latest teachings of science, and represents the spirit as well as the letter of the laws that require these truths taught all pupils in all schools. We are therefore happy to indorse and commend the same for use in schools.

MARY H. HUNT,

National and International Superintendent
Department of Scientific Instruction of the
Woman's Christian Temperance Union.

ALBERT H. PLUMB, D. D.
DANIEL DORCHESTER, D. D.
HON. WILLIAM E. SHELDON. } *Advisory board for the United States.*
REV. JOSEPH COOK.

COPYRIGHT, 1884, 1886, 1889,
BY D. APPLETON AND COMPANY.

T. OUTLINES.

Printed by
D. Appleton & Company
New York, U. S. A.

PREFACE.

IT has been my aim in preparing this volume to compress within the narrowest space such a clear and intelligible account of the structures, activities, and care of the human system as is essential for the purposes of general education. I have also sought to present the facts and principles of the subject in such a natural order as will best subserve the true ends of scientific education. Useful books of information upon physiology are already numerous, but they are too generally deficient in making the science valuable as a means of mental training. Of course, the great object of physiology is to teach how to preserve health, but this is not best done by merely memorizing rules. The rules must be supported by reasons, and if there is not some thorough understanding of the mechanism and powers of the human body, such as will task the efforts of the student, the real fruits of knowledge will not be gained. I have accordingly given prominence to the anatomical and physiological facts which are necessary preliminaries to instruction in hygiene, and in the reasonings upon these facts I have aimed to attract and interest the pupil, to teach him something of the scientific methods of approaching the subject, and to connect new acquisitions logically with those already gained, so that the knowledge of the subject may become, as it were, organized into faculty in the minds of the students. So important has it seemed to me to impress deeply upon the pupil's mind the laws of connection and dependence among the various parts of the living system that I have thought it best to present this view, in outline, at the very outset. I have, therefore, prefixed to the volume a General Analysis, which, while it serves as a table of contents, is interspersed with running comments explaining the general relations of the different organs and processes, and I recommend that this analysis be carefully read by the pupil, so that he may become familiar with its argument before proceeding to the detailed study of the text.

One of the greatest modern reforms in scientific study is undoubtedly that which makes it more and more objective, so that the student shall constantly confirm the knowledge he gets from the book by reference, as far as possible, to the objects themselves, making his acquaintance with them direct, and his information real. The various sciences lend themselves to this mode of study in different degrees; chemistry and physics favoring experiment, and botany offering systematic observation more than any other scientific subjects. Physiology is less favorable to the objective method. For the purpose of ordinary education, it must be chiefly taught from the book, with such accompaniments of lectures and illustrations by charts as the circumstances will allow. But even here much may ·e done to give the pupil more correct ideas of the

elements of the subject than can be obtained from
the book alone. A good manikin is an invaluable
help to the popular study of anatomy and physiology. Human dissection being out of the question,
the manikin, which can be taken to pieces so as to
show all the organs in their situations, connections,
and relative dimensions, will afford the pupil a vivid
and exact conception of the dependent parts of the
living body, and make his physiological knowledge
truthful and actual. A manikin for school purposes costs about $250, and may be imported from
Paris,* where they are made, free of duty for educational institutions.

A great deal is also to be learned from such
rough dissections of organic tissue and structures
as may be made anywhere. Every butcher's shop
is full of specimens of all parts of animals, that can
be cheaply obtained for examination, and parents
and teachers should encourage pupils to make such
rude dissections as are practicable, and will help to
give correct ideas of the relations and functions of
the different organs.

The study of the minuter parts of organized
beings with the microscope, histology as it is called,
has come into great prominence in modern times,
and may be said to have revolutionized the science
of life. No class in physiology should be without a
microscope for the direct study of cell-structures
and the finer tissues of both plants and animals. A
suitable instrument, with a magnifying power of
three hundred and fifty diameters, will show the
circulation in the web of a frog's foot, and open a
new world of fascinating and wonderful observation,

* Auzoux is the principal manufacturer of these models.

while it may be bought for sixteen dollars. Microscopic preparations of blood-corpuscles, muscular and nervous tissues, and sections of organs may be got for about twenty cents apiece, but it is desirable that the pupil should not rely upon these, but should learn the method of preparing and mounting objects himself. The microscope is not to be recommended as a mere toy to amuse idle curiosity; there is work connected with it which is in a high degree educational. It cultivates critical observation and careful manipulation, and is invaluable as a means of self-education. The little hand-book of Phin will be found useful in guiding beginners with this instrument.

In compliance with the laws of many States requiring the study in schools of "physiology and hygiene with special reference to the nature of alcoholic drinks and other narcotics, and their effects upon the human system," much additional matter relating to that subject will be found in this book. In its preparation the author has had the valuable co-operation of Mrs. Mary H. Hunt, Superintendent of the National and International Department of Scientific Instruction of the Woman's Christian Temperance Union, who has kindly gone over the entire book, and whose familiarity with the subject and with the laws relating thereto has given especial importance to her suggestions regarding the selection and arrangement of the new matter.

The illustrations are largely copied from Gray's "Anatomy," though I am also indebted to Dalton's "Physiology," to Flint's "Physiology," to Ranney's "Applied Anatomy of the Nervous System," to

Rüdinger's " Topographisch-Chirurgische Anato-mie," and to Neumann's " Hand-Book of Skin Dis-eases." Many of the figures I have altered to suit my purpose, and the necessary descriptions are so inscribed upon or near them as to do away with the inconvenience of lettered references. A few of the cuts are original.

For the use of material other than the illustra-tions, I have to acknowledge my indebtedness to Flint, Beaumont, Stricker, Neumann, Rüdinger, Paget, Maudsley, Reynolds, Aitken, Huxley, Soel-berg Wells, Uhle and Wagner, Foster, and espe-cially to Dalton.

R. S. T.

GENERAL ANALYSIS.

PART I.—INTRODUCTION.

Gives certain necessary definitions, and describes the cell and its properties as being the real basis of all more fully developed living organisms.

PART II.—ORGANS OF MOTION.

A large body entirely composed of cells would be a soft, jelly-like mass, incapable of locomotion or of self-protection. But to obtain food it must be able to move from place to place, and also to move its different parts with reference to one another. For these purposes there must be points of resistance and points of support. These points are furnished by the bones, which act as levers, the joints being the fulcra.

But levers alone are of no use. The bones form a strong framework for the body, but they can not move themselves. To produce motion, organs are required, which can become longer or shorter, under varying circumstances. Such organs are the muscles.

PART III.—ORGANS OF REPAIR.

Energy is never lost or created. If the body loses energy in one way, it must gain it in another, or it will soon be worn out. Every muscular contraction wastes a certain amount of material, and an equal amount must be again supplied. This is done in the form of food.

But food, as it exists outside of the body, can not be appropriated by the wasting tissues. It must first be prepared. The process of preparation is called digestion.

After the food has been so far prepared, it must in some way be carried through the body to all its different parts, that each may take what it requires for its sustenance. This is accomplished by means of a fluid which passes continually and rapidly through all parts of the body, carrying the nutritious material. This fluid is the blood.

But the blood, besides carrying nutriment, must also remove the waste and used-up matters, which injure the health if they remain in the body. There is also a gas, called oxygen, which is found to be necessary to the processes of nutrition. This gas exists in the air, and is taken from the air into the blood. The process, by which the blood gets rid of impurities and takes in oxygen, is called respiration.

The blood can not visit the different parts of the body of its own accord. It is a fluid, and must be propelled. There are organs for this purpose, which keep up what is called the circulation of the blood.

PART IV.—ORGANS OF CO-ORDINATION.

The motions of the body, the continual waste and supply, and the processes of digestion and circulation, form a very complicated series of phenomena. Certain parts of the body require more blood at certain times than at others. Processes taking place at the same time in different parts of the body might conflict and interfere with each other. We find, therefore, a system of organs whose function it is to harmonize or co-ordinate all these processes, to produce a sympathy between them, and make them all work together for the common interest. This is the nervous system.

PART V.—ORGANS OF PROTECTION.

All the organs previously described form a very delicate structure, which is continually exposed to external injurious influences. It is exposed to heat and cold, to blows and scratches, and all manner of violence, and so we find it enwrapped in a strong covering, which protects it from these influences, partly by its own strength and toughness, and partly by certain organs which are imbedded in it, and form a part of it. This organ is the skin, with the various glands and other structures found therein.

PART VI.—ORGANS OF PERCEPTION.

The body being now complete, so far as its movements, nutrition, co-ordination of parts, and protection are concerned, we see that, as its food must come from outside, there must be organs to bring it into relation with the external world, to aid it in its search for food, and to protect it during the search. These organs are the organs of the senses, which bring us into relation with what is outside of us, and in this way are the sources of our ideas. The elementary one of these senses is touch, the others being only modifications of it.

The body being now practically complete, we find still another organ, whose function it is to enable us to communicate our ideas to others. This organ is the larynx, the organ of speech, that wonderful faculty which has had so much to do with creating the tremendous gap that exists between man and the lower animals.

PART I.

INTRODUCTION.

CHAPTER I.

1. Definitions.—The science which tells us about the different parts of the body, what they are, where they are, and how they look, is called *anatomy*.

The science which tells us about the purpose of these parts, what they do and how they do it, is called *physiology*.

The science which tells us what will interfere with the working of these parts, what will injure and what will help them, and how to avoid what is hurtful, is called *hygiene*.

A part of the body, which is so small that when it has been separated from other parts it can not be further subdivided without the destruction of its organization, is called an *anatomical element*, as a cell or a fiber.

Two or more anatomical elements, united or interwoven so as to form one substance, make what is called *tissue ;* e. g., fatty tissue, connective tissue, etc.

A part of the body, which is made up of anatomical elements and tissues, together forming one whole, which can be separated from the rest of the body as an entire mass, and which performs a particular function, is called an *organ;* as, the liver, the heart, a bone, or a muscle.

A number of organs, similar in structure, but differing in size and shape, and spread throughout the body, are called a *system;* as, the nervous system, the arterial system, etc.

Several organs, which differ in structure but are so connected as to work together for a particular end, are called an *apparatus;* thus, the stomach, liver, etc., constitute together the digestive apparatus.

· The work that is done by a healthy organ in the body is called its *function;* e. g., the secretion of bile is a function of the liver.

Any substance whose nature it is when absorbed into the blood to injure health, or destroy life, is called a *poison.*

A *narcotic* is something that, when introduced into a living organism, deadens its sensitiveness, and so diminishes its ability to perform its natural functions.

When a poison does its injurious work wholly or in part by paralyzing or deadening the nervous system, it is called a *narcotic poison.*

As narcotic poisons will be frequently referred to in the following pages, a brief description of those best known is given here.

Opium, a powerful narcotic, is a blackish or brownish gummy substance obtained by evaporating the milky juice of the poppy. Its composition is very complex, one of the principal substances it contains being morphia (or morphine), usually forming about ten per cent of its weight. *Laudanum* is an alcoholic tincture of opium, twenty-five drops being about the equivalent of one grain of the drug. *Paregoric* is a compound tincture sweetened with honey and flavored with anise, containing about one grain of opium in each tablespoonful.

Alcohol, a product of fermentation, is a narcotic poison. Its great affinity for water, and its power of coagulating albumen, render it destructive of living tissues with which it is brought in contact. It is a colorless, transparent fluid, of light specific gravity (0.7938), is very inflammable, burning with a smokeless blue flame, and has been frozen at the exceedingly low temperature of −203° Fahr., when it solidifies into a white mass. Pure alcohol is very difficult to obtain, on account of its strong affinity for water. It is composed of carbon, hydrogen, and oxygen (C_2H_6O).

There are various kinds of alcohol—that which imparts the intoxicating quality to wines, etc., being known as ethyl alcohol. Other forms are found in rotting wood, in decaying grains, etc. For further account of the origin and nature of this substance and of drinks containing it, see Part III, Chapter II.

Tobacco is the dried leaf of a plant called by botanists *Nicotiana Tabacum.*

The tobacco-leaf is cured in various ways, and put through manufacturing processes which affect its quality. Like opium, it is a very complex substance, but the active ingredient, which renders the use of it so attractive to many people, is called nicotine. This is a volatile, colorless liquid of a specific gravity of 1.018, which turns brown on exposure to the air. It constitutes from one to nine per cent of the weight of the tobacco, the amount varying greatly in different specimens. Nicotine is a deadly poison like Prussic acid, destroying life in small doses with great rapidity.

Chloroform ($CHCl_3$) is a narcotic poison, obtained by distilling chloride of lime, water, and alcohol together. It is a thin, colorless liquid (specific gravity, 1.525), of agreeable odor and sweetish taste. When inhaled, the vapor produces temporary insensibility.

Chloral (CCl_3CHO) is made by passing chlorine gas into ethyl alcohol. It is a thin, oily, colorless liquid, of a pungent odor, but almost tasteless (specific gravity, 1.502). It is used by physicians to induce sleep.

2. Minute Structure of the Body.—The body, when its parts are examined with the microscope, is found to be made up mainly of *cells*, *fibers*, and *fluids*. The cell is considered to be the original element out of which every other element in the body is formed; fibers, fluids, etc., being derived from or generated by previously existing cells.

The different consistency of different organs is due to the varying proportions of these elements. If the fibers are in the largest proportion, the tissue or organ will be hard, tough, and elastic; if the cells form the greatest part, it will be soft, inelastic, and yielding.

3. The Fiber.— The *fiber* proper (Fig. 1) is a slender thread, composed of a hard whitish or yellowish substance, sometimes elastic and sometimes not, but very tough

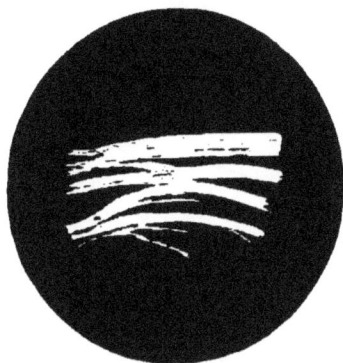

FIG. 1.—Fibrous tissue.

and strong. Fibers are found in almost all parts of the body, binding the parts of organs together, and

2

constituting almost the entire mass of some parts, as the tendons or sinews, for instance. The word "fiber" is also used of certain portions of muscular and nervous tissue, in a different sense from the one given above. These differences will be explained hereafter.

4. **The Cell.**—The cell is the most important structure in the living body, whether animal or vegetable. Life resides in the cell; and every plant or animal may really be looked upon as a mass composed of billions of cells, each of which is alive, and each of which has its own part to play in the nourishment of itself and the rest of the body.

A single cell * (Figs. 2 and 3) is a portion of al-

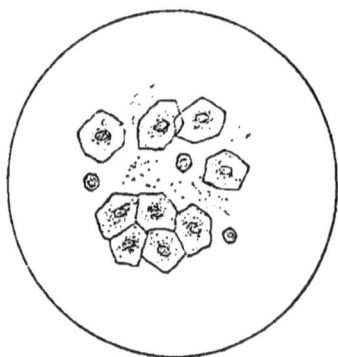

FIG. 2.—Epithelium from the mouth.   FIG. 3.—Liver-cells.

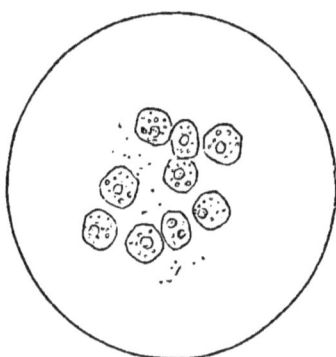

buminous matter, which has by some been called protoplasm,† sometimes surrounded by a thin mem-

* Scrape gently the surface of the tongue, and put the fluid thus obtained under the microscope. Plenty of cells will then be seen, similar in appearance to those shown in Fig. 2.

† *Prō'toplasm*, from two Greek words, signifying the first (or primitive) formed matter, because, so far as we know at present, it is the

brane and sometimes not, and usually having in its interior what looks like a smaller cell. This small body is called the *nu'cleus* of the larger one. Inside of the nucleus is often found another exceedingly minute body, or sometimes a mere shining point, called the *nucle'olus* (see Fig. 52).

5. Protoplasm. — The protoplasm, or matter which forms the mass of a cell, is of a semi-fluid consistency, somewhat like jelly, and can not be distinguished by chemical tests from albumen. Hence it is said to be albuminous in its nature, resembling to some extent the white of a raw egg, which is almost pure albumen. All cells are so exceedingly small, being rarely more than $\frac{1}{1000}$ of an inch in diameter, that we really know very little of their minute structure, on account of the difficulty of investigating them with such imperfect instruments as we have.

6. Granular Matter.—The points just mentioned are the most characteristic of the cell. Besides the cells, fibers, and fluids, there is a great deal of matter in different parts of the body, which has been formed or deposited by the cells at different periods of their growth. This matter, under the microscope, sometimes appears granular, or as if made up of very minute specks, and sometimes almost transparent. Some of it is found to be fat in a finely-divided state, but some of it is albuminous, and some of it contains mineral matter in considerable amount.

7. Difference between Living and Dead Cells.— A living and a dead cell look precisely alike, except-

simplest form of living matter, and makes up the only part of all animal and vegetable bodies which shows signs of life.

ing that the dead one is motionless. A living cell, minute as it is, frequently possesses the power of independent motion, or at least is able to change its form (Fig. 4). It also can take up nourishment into its mass, and can propagate itself. The movements of the cell can be beautifully seen in the white cells of the blood, which will be described hereafter.

8. Growth of Cells.— A cell propagates itself in several ways; one of the most common is by dividing into two parts. This operation has been often watched under the microscope by skilled observers. The change is seen to begin in the nucleus, and, as that divides by a line through its center, the protoplasm of the cell arranges itself in halves around each part of the nucleus, its surface dipping in toward the center, until finally the approaching surfaces meet and the

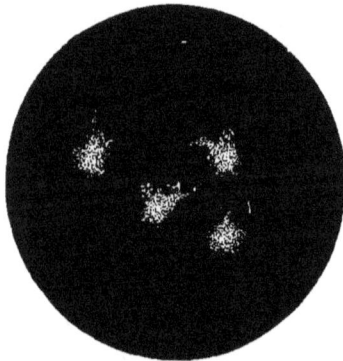

FIG. 4.—White cells in motion.

FIG. 5.—Cell dividing and forming two new cells.

cell is divided into two new cells, each with its nucleus, and in every way complete (Fig. 5). This division goes on with great rapidity. The secretion from the throat and nose in nasal catarrh is composed mainly of cells, which are thrown off by millions during an inflammation.

9. **Other Properties of Cells.**—Cells also possess other powers which enable them to perform their important offices in the body. They are able to select certain substances out of a general mixture, and reject others. This is done by the liver-cells, for example, which secrete the bile, and by the cells of those glands which secrete the saliva. The cells of the brain act, in some unexplained way, as the instruments of thought. The cells in the kidneys separate matters from the blood which are very poisonous, and have to be expelled from the system.

The power of division and of numerical increase of cells is not unlimited. If a portion of the body is wounded, it is healed again by the active efforts of the uninjured cells in the borders of the wound. The action of these cells ceases, however (if the part is healthy), to reproduce substance, when the part made vacant by the injury has been filled up. Why does this action of the cells, once started, not continue until the body dies? Why does the replacement of tissue cease as soon as the former surface is reached? This question can not at present be answered.*

Thus we see that the cell, minute as it is and simple as it is, performs its office in the body with

---

* When the cells in the borders of a wound or sore are not in a healthy condition, they sometimes increase in number very rapidly, but the new cells, instead of being like the older ones, are larger, chiefly owing to the greater amount of fluid in their interior. This makes them soft and spongy, and seems to interfere with their functions. They do not nourish themselves properly, and they increase and grow beyond the limit where they should stop, and where, if they were healthy, they would stop. This unhealthy growth is what is known as *proud flesh,* and it has to be repressed by proper surgical treatment.

care and evidence of forethought and intention. It does what is necessary and no more. It does not act blindly. It does all it does with a purpose. Where and what is the intelligence that directs the active living cell to repair so far and no farther, to eat this and reject that, to multiply up to a certain point and then stop, and, most wonderful of all, to take upon itself the duties of other cells when they are sick and unable to act,* and stop performing these extra duties, when the other cells recover?

10. Effect of Alcohol and Tobacco on Living Cells.—The cells which compose the mass of the body, being very delicate in their structure, are easily acted upon by whatever comes in contact with them. If other substances besides those contained in our natural food and drink are introduced into the body, the cells are likely to be injuriously affected; and this is especially true of narcotics, of which alcohol and tobacco are the most universally used. These substances blunt the sensitiveness of these cells, retarding the changes in their interior, hindering their appropriation of food and elimination of waste matters, and therefore preventing their proper development and growth. The more recently formed such cells are, the less are their powers of resistance; so that in young persons who are growing the effect of such drugs is especially potent and harmful.

---

* When the kidneys, for instance, are diseased, so that the excretion of urine is interfered with, it is found that some of the poisonous matters which usually pass out through them are ejected from the body through the lungs and skin. In such cases, physicians try to assist this process by inducing active perspiration, so as to relieve the kidneys from work as much as possible, and allow them to rest until they get well.

PART II.

ORGANS OF MOTION.

CHAPTER I.

BONES.—GENERAL STRUCTURE.

11. Uses of the Bones.—If the body were composed merely of cells, such as have been described, with fibers and fluids, it would be a shapeless, jelly-like mass, incapable of locomotion, and of self-protection. There is a necessity, in such a large mass, of points of support and resistance, and the organs or tissues which furnish such points must be tough, hard, and elastic. Such organs are the bones, which form the framework of the body and determine its shape and size. Their most important offices are two in number, viz., to act as levers and points of support and action for the muscular parts, and to protect the soft and delicate organs from external injury.

12. Living and Dead Bone. — A bone, as we usually see it outside the body, is as different from a living bone as the skin of a corpse is from the living skin. We usually see it deprived of blood, while in the living body it is full of it, and is of a pinkish-white color externally, and deep red within.

13. Composition of Bone.—To accomplish the

two purposes above mentioned, bones must be hard
and tough, in order to maintain their stiffness when
the muscles pull upon them, and also to be able to
resist external blows. They must also be in some
degree elastic, or they would be too brittle for use,
and would snap in two under great pressure. Accordingly, we find all bones composed of two kinds
of material, so thoroughly mingled and united that,
when either kind is removed, the bone still retains
its peculiar shape and size, although of course it
does not weigh as much as before. About two
thirds of the weight of every bone in the adult consists of earthy substances, mostly lime phosphate
and lime carbonate, and the remaining third of animal matter, part of which can be separated from
the rest of the bone by long boiling, and is known
as gelatine.

If a bone be burned in a hot fire, all of the animal matter will be destroyed, and the earthy matters left. These will still retain the shape of the
bone, but will be white in color, and will easily
break and crumble in the fingers. If a bone, on
the other hand, be soaked for a time in dilute hydrochloric acid, all the earthy matter will be dissolved out, and the animal portion left. This, as in
the other case, will retain the shape of the bone,
but will be flexible and tough, and may even be
tied in a knot.

By the combination of these two kinds of matter, then, the bone is made hard, tough, and elastic,
and admirably adapted to its uses in the body.

14. The Composition of Bone varies with Age.
—In infants and children, the amount of animal matter in the bones is proportionately greater than in

the adult, and so the bones of very young people will often bend when injured, instead of breaking. Surgeons call this the "green-stick" fracture, be- cause the bone is bent like a green twig, only a small portion of it on the outside of the bend being broken or torn apart. As a person grows older, the amount of earthy matter increases, until in old people the bones become very brittle, and break with very slight blows.

15. Varieties of Bone and their Structure.— Bones are divided, according to their shape, into *long* bones, *short* bones, *flat* bones, and a fourth kind, called *irregular*, which combine qualities belonging to the other classes. The *long* bones are found only in the limbs, and are the most important to the sur- geon, as it is in them that most fractures and other injuries occur. They are divided into a *shaft* and *extremities*. The *shaft* of every long bone consists of hard, compact, closely-grained tissue, somewhat like ivory. This is the only part used in the man- ufacture of ornaments, buttons, knife-handles, etc. The *extremities* of these bones form the *joints*, and, in order to give greater security and a greater purchase to the muscles as well as a greater surface for their attachment, the ends are much larger than the shaft. The tissue of which they are composed is also not so hard and close in texture, as, if it were so, the bone would be too heavy. There is no finer example of economy of material and the combina- tion of strength with lightness than the structure of the long bones (Fig. 6). The ends are made of fine threads of bone interlaced and crossing and supporting each other, so as to make a sort of *spongy* tissue, full of little cavities, and yet very strong and tough. And even the shaft of the bone is not solid, but, as every one knows, is hollow in the middle. This hollow space and the little cavities of the ends of the bone are filled with what is called *marrow*, a substance composed chiefly of blood-vessels and fat, which has important duties to perform in the growth and nourishment of the bone. The other varieties of bone are composed entirely of the *spongy* (or cancellous) tissue, with a thin layer of hard, compact tissue on the surface.

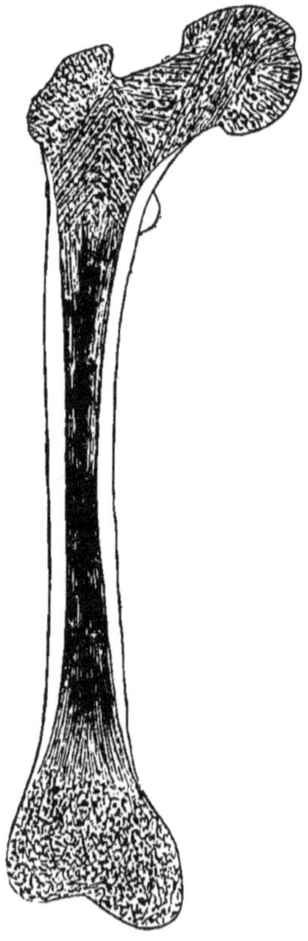

16. The Periosteum and the Minute Structure of Bone.—All the bones are covered with a very tough, strong, fibrous membrane, called the *perios'teum*, excepting at the parts which enter into the formation of the joints, where they are covered with cartilage. This membrane adheres so closely to the bone as to require considerable force for its separation. It seems to form a part of the bone. Now, the periosteum and the marrow of the bones are necessary to their growth and nourishment. The

FIG. 6.—The right femur, or thigh-bone, sawn in two lengthwise. Notice the arrangement of the bony fibers at the upper end, its peculiarity being somewhat exaggerated so as to make it more plain.

blood-vessels and nerves spread and divide in these tissues before entering the actual substance of the bone. The bone itself is full of minute channels and tubes varying in size from $\frac{1}{200}$ to the $\frac{1}{20000}$ of an inch in diameter, through which the blood circulates, and the smallest of these tubes are connected at one end with exceedingly minute cavities in the bone, in each of which lies a little cell, which does the work of nourishing, repairing, and enlarging the bone (Fig. 7). Thus we see that, even in so hard and firm a tissue as bone, what has been said about cells holds true. They are the real life of the bone; they separate from the blood the necessary material and deposit it around themselves, somewhat as a crab renews his shell every year after getting rid of the old one.

Fig. 7.—Cross-section of bone, magnified. The small black spots are the cavities in which the bone-cells live. The fine lines are canals through which the plasma (section 122) of the blood passes. The large holes are for blood-vessels.

17. Uses of the Periosteum.—It has long been known that, when the periosteum is severely bruised and separated from the bone by violence, the portion of bone deprived of the periosteum dies and has to be removed from the body. It is also found that a portion of bone, or even an entire bone, may be removed from the body, and if it be carefully done, so as to leave the periosteum in its place, the bone will grow again. A remarkable example of this was a case operated upon by the late Dr. James R. Wood, of New York. In a young woman, whose lower jawbone had become dead and caused her great suffering, this distinguished surgeon removed the whole jaw, leaving the periosteum and even the teeth, held in their places by an apparatus made for the purpose. The entire bone grew again, and the teeth became fixed in it as it grew. The person died several years afterward, and her skull, showing the result of this wonderful operation, is in the museum at Bellevue Hospital.*

* Other experiments have even shown that, if a piece of fresh living periosteum be transplanted from a bone to a muscle, it will produce bone in its new situation. These remarkable qualities of the periosteum have been explained by some, by supposing that, in each case of operation or experiment, some of the minute bone-cells have adhered to the periosteum when the mass of the bone was removed, and that they were the chief agents in forming the new bone.

CHAPTER II.

18. **The number of bones** in the human body is two hundred (Fig. 8). At one period of life they are nearly all cartilaginous, but the cartilage is gradually changed into bone. This process of change, *ossification*, as it is called, is not complete before the twenty-fifth year of life, and therefore no person can be called really grown up until that time.

19. **The Vertebræ.**—The foundation, so to speak, of the body—that portion of the skeleton to which the remainder is attached, and from which it is built up—is the *spine*, or backbone (Fig. 9). This is composed of many small bones, all of the same general pattern, called *ver'tebræ*. The principal part of the *vertebra* (Fig. 10), called the *body*, is shaped very much like a wooden pill-box slightly hollowed out on the top and bottom. The bodies of the vertebræ form the front of the spinal column. From the rear of each of these bodies are offshoots of bone, which unite in such a way as to leave a hole about half an inch in diameter running up and down. These vertebræ are placed one above another, with elastic *pads of cartilage* between their bodies. These pads are so thick that, taken all together, they make up about one fourth of the whole

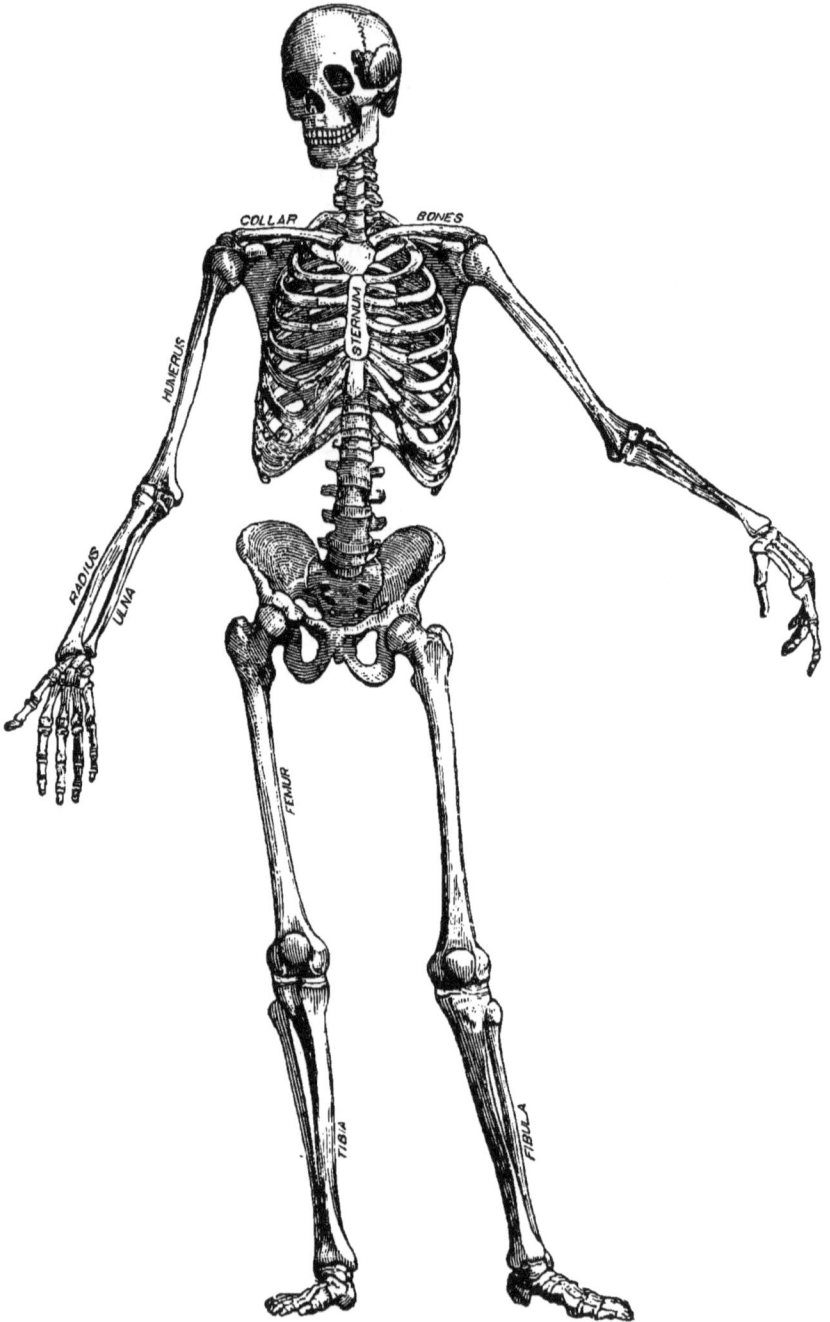

COLLAR BONES

HUMERUS

STERNUM

RADIUS

ULNA

FEMUR

TIBIA

FIBULA

FIG. 8.—The skeleton.

length of the backbone. The vertebræ being applied to each other in this way, it is evident that the holes just mentioned, which are surrounded by bone, will form a continuous canal (the spinal canal) running from the skull down the back. This canal contains the *spinal cord*, which, next to the brain, is the most important part of the nervous system. At the sides of the spine, throughout its whole length, are holes, out of which pass nerves supplying the muscles and other organs of the body, and into which pass the blood-vessels that nourish the spinal cord.

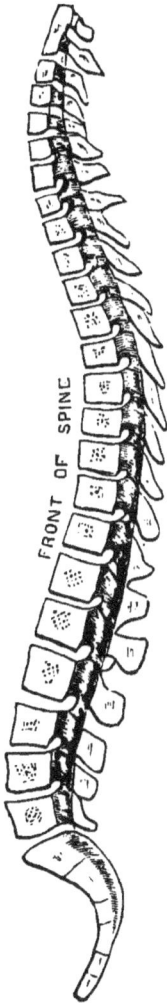

20. The Spine. — The backbone, being composed of so many pieces, is very movable. The power of motion varies, however, in different parts. It is greatest in the neck, and least in that portion of the back to which the ribs are attached. In the human being, the neck is not so flexible as in many animals. Birds, in particular, can look directly backward.

Notwithstanding the great power of motion in the spine, the different bones are very strongly united and protected by powerful ligaments and muscles, which

FIG. 9.—The spine, sawn in two lengthwise, showing the spinal canal and the holes between the vertebræ, where nerves and blood-vessels pass out.

render it difficult for a vertebra to slip out of place, and such an accident is one of the rarest which a surgeon is ever called upon to treat.

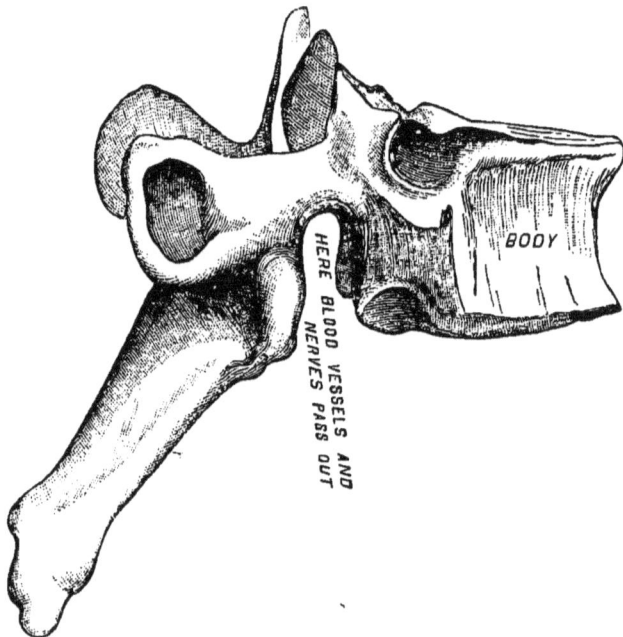

FIG. 10.—A vertebra.

The elastic pads between the vertebræ deaden all shocks of the body and prevent them from injuring the brain. These pads become compressed during the day, especially when a person is much on his feet, so that at night-time the body is from one quarter to one half an inch shorter than it is in the morning. During sleep or rest the elasticity of the pads causes them to resume their original shape and thickness.

21. The Skull.—The *skull* is the only portion of the skeleton whose principal office is the protection of soft parts within it. Accordingly, we find

that its bones are differently composed and differently put together from the other bones in the
body. Those forming the outside of the skull, immediately surrounding the brain, and most exposed
to blows, are composed of three layers. The outside layer is the thickest, and is tough and somewhat elastic. The innermost layer is very thin, but
hard and brittle, so that it is called the *vit'reous*
(glassy) table of the skull. Between these layers is
spongy tissue, like what has been before described.
This deadens every blow upon the head, and the
safety of the brain is still further provided for by
the arched shape of the skull, which tends to dif-
fuse the force of a blow. The protection afforded
by the shape and structure of the outside portion of the skull is very great, and it is a well-
known fact in surgery that a blow upon the top
of the head, without breaking the bone on which
it falls, may break the bones at the base of the
skull, immediately opposite the spot of the blow,
by the mere force of the shock, although the latter
bones are much thicker and more massive than the
others.

There is only one movable bone in the skull and
that is the *lower jaw.* If the upper jaw be made to
move in eating or speaking, it is only by moving
the whole head where it joins the neck.

22. Sutures of the Skull.—The bones of the
skull are joined together by what are called *sutures*
—i. e., their edges are jagged and irregular, and fit
together like dovetailed boards (Fig. 11). This
renders the arch of the skull more compact, and,
as far as resistance to pressure is concerned, the
bones might be considered as one piece, while the

interruptions at the sutures tend to deaden the shock of a blow.

23. The Frontal Sinuses.—In the front of the skull there are two cavities in the substance of the bone itself. These are situated just above the eye-brows, and are called the *frontal si'nuses* (Fig. 12). The layer of bone which forms their front wall causes the prominences just over the eye-brows, and, as the cavities increase in size with age, this portion of the forehead becomes more prominent. The cavities are lined with mucous

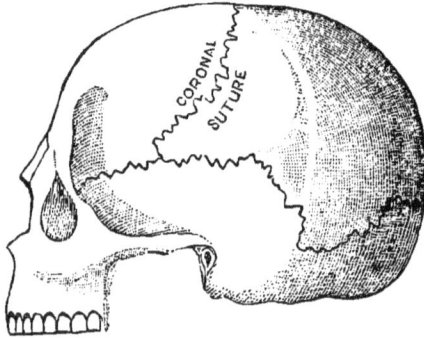

FIG. 11.—General outlines of the skull, showing the sutures.

membrane, and are connected with the inside of the nose by a canal or small passage, so that, when a person has a severe cold in the head, the inflammation sometimes runs up this passage into the frontal sinuses. When this is the case, the person has a dull, stuffy headache in that locality, due to the swelling of the mucous membrane.

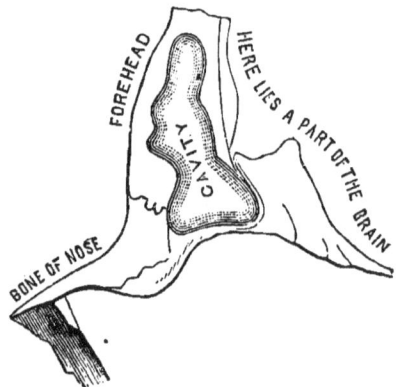

FIG. 12.—Frontal sinus.

24. The Ribs.—The bony part of the walls of

the chest is made up of *twenty-four ribs* and the *breast-bone*, together with part of the spine behind. There are twelve ribs on each side, the first, nearest the neck, being usually the shortest. They increase in length from the first to the seventh, and then diminish, so that the twelfth is also quite short. They are flat and narrow, and are attached at one end to the spine, in such a manner that they move easily up and down, while the other end is attached to the breastbone, or sternum, by means of a piece of cartilage, varying in length with the length of the rib. The eleventh and twelfth ribs are not attached to anything at their forward end, and hence are called floating ribs.

The ribs are attached to the spine in such a way that all of them move together up and down. In front, the stiff but elastic cartilage allows motion in every direction. Now the shape of the ribs is so peculiar, being a sort of double curve, that when they are raised at the sides, the ends which join the breastbone are pushed forward, and of course carry the breastbone with them. So it is evident that at every inspiration the diameter of the chest increases from front to rear as well as from side to side.

25. Natural Shape of the Chest.—In young people the *cartilages* are soft, but they grow harder as age advances, and become partially turned into bone. In youth they yield to pressure to such an extent that by tight lacing the shape of the chest is sometimes made exactly the reverse of what it ought to be (Fig. 13). The ribs naturally form a cone, with the smaller end uppermost, but it is not uncommon to see the smaller end at the waist. Nature will endure a great deal of meddling, but it is not always safe to trifle with her, and all persons who carry tight lacing too far will inevitably suffer.

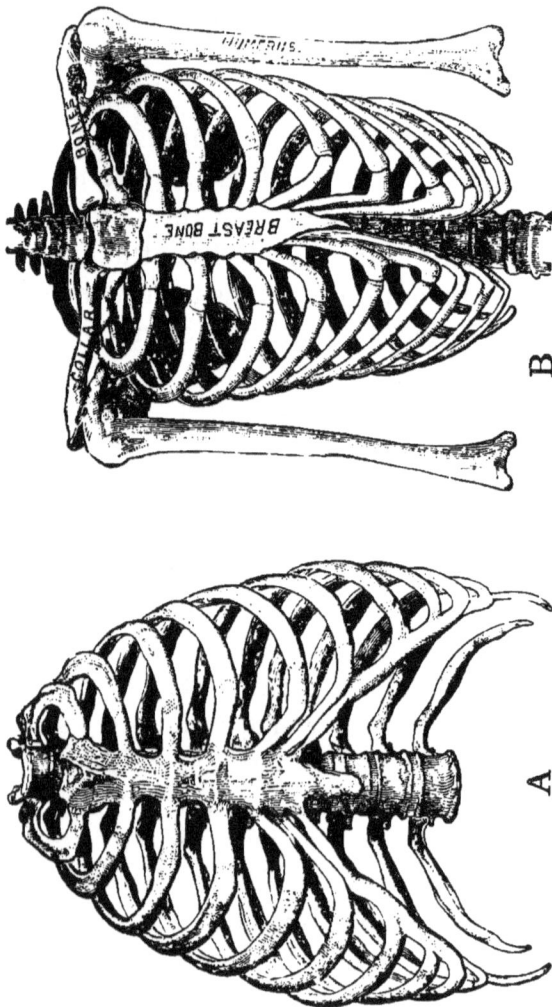

FIG. 13.—A represents the normal appearance of the ribs. B represents part of a photograph of the skeleton of a young woman of twenty-three years, showing the distortion of the ribs produced by tight lacing in an actual case.

26. The Limbs.—More than one half of the bones in the body are found in the *limbs*. Out of two hundred bones, they contain one hundred and twenty-six, and these are so constructed and so arranged as to afford a great variety of movement. The upper and lower limbs are what is called *homol'ogous* in their parts—i. e., each bone in the arm has its counterpart in the leg, with only slight apparent exceptions. Thus, the shoulder - blade corresponds to the body of the hip-bone, the collar-bone to the front of the hip, the arms to the thigh, the two bones of the fore-arm to the two of the leg, the wrist to the ankle, and the hand and fingers to the foot and toes. The similarity and correspondence of these parts are quite clear in the skeleton.

27. The Joints.— To render move- ments possible, the skeleton is broken up in its whole extent by numerous *joints*. The surfaces of the joint are not covered by periosteum, but by a firm, bluish- white, smooth, and very elastic sub- stance called *carti- lage*.

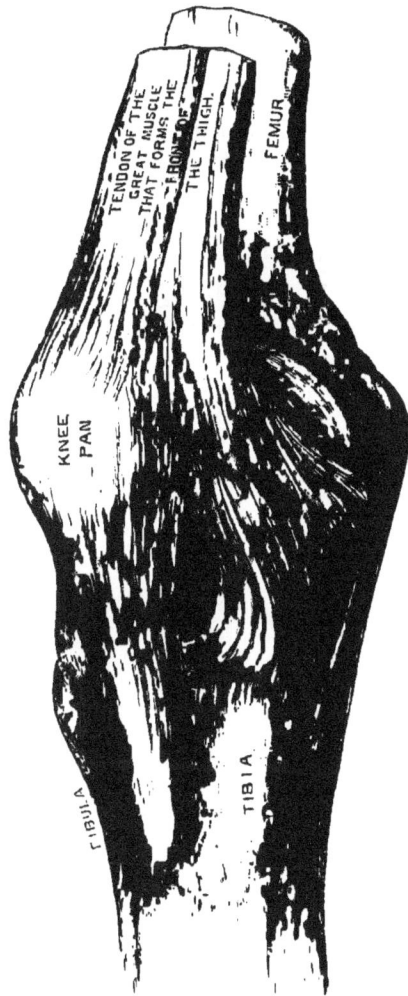

FIG. 14.—The right knee-joint, showing how strongly it is bound about by ligaments.

The two cartilage-covered surfaces in every

joint are in contact with each other. The joint is closed entirely by the *syno'vial membrane,* which passes over from one bone to the other, all round the outside. This membrane is exceedingly smooth and delicate, and its inner surface exudes a fluid very much like the white of an egg, which moistens the surface of the joint and renders every movement easy and frictionless. Outside of these structures are *ligaments* (Fig. 14), which hold the bones firmly in their places. Ligaments are composed of white, fibrous tissue (Fig. 1)—i. e., of tough, inelastic fibers or threads running parallel with each other, of a shining, silvery-white color. They are flexible, so as to allow of considerable lateral movement, but are tough and exceedingly strong, so that they hold the ends of the bones close together. Thus, the construction of the joints is such that they are strong, flexible, elastic, and supple.

CHAPTER III.

28. Injuries in General.—The injuries to which the bones are most liable are *fractures* and *dislocations*. If the bone be fractured, the jagged ends of the broken bone irritate the parts about them, and the muscles contracting pull the broken ends out of their proper relation to each other (Fig. 15). In the dislocation, the end of the bone is out of its proper place. But the limb is movable at the point of fracture, while it is always stiff and fixed at the point of dislocation. In a fracture, also, the ends of the bone, if gently moved against each other, produce a peculiar grating feeling, which always tells the surgeon with certainty that the limb is broken.

29. Fractures.—Bones are rarely broken straight across, excepting in very young persons. The fracture is usually oblique, and so the broken ends slide past each other, and the limb is shorter than it was before the accident. In a broken thigh, the bone is surrounded by such a thick mass of muscles that, even if the broken ends are pulled by force into their proper places, it is impossible to keep them there. They will always slide past each other to a slight extent, and a person never recovers from such an accident, without having the injured limb from half an inch to an inch shorter than the sound one. This has been shown and proved by thousands of careful measurements, and should always be borne in mind when there is a temptation to blame a surgeon for fancied neglect.

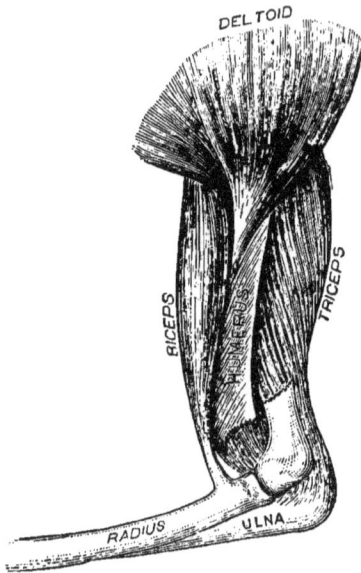

When a fracture occurs near a joint it is a much more serious accident, for the inflammation which follows the injury involves the parts about the joint, and sometimes the joint itself, which may be left stiff and almost useless for a long time afterward. This is particularly the case with fractures near the wrist, for the slow recovery of which the surgeon is so often blamed.

FIG. 15.—Fractured humerus, showing how the muscles pull the ends of the broken bone out of place.

30. Dislocations. — When a bone is dislocated there is always a certain amount of injury to the parts about the joint. The ends of the bone are so carefully and strongly guarded and fastened by ligaments and muscles, that these must necessarily be considerably torn and bruised, in order to let the bone out of its place. Thus it happens that a dislocation often gives rise to more pain and suffering immediately after the accident than a fracture.*

* Sometimes the violence resulting from a fall is not sufficient either to break or dislocate a bone, and yet the parts about a joint are so se-

The vast majority of dislocations occur in the shoulder and hip joints, and are usually caused by a blow on the end of the bone when the limb is firmly extended, as when a person is falling and tries to save himself by stretching out his hand. The lower jaw is sometimes dislocated, and then the mouth remains wide open until the dislocation is reduced, rendering the sufferer a somewhat ludicrous as well as pitiable sight. This accident has been known to occur during a prolonged yawn.

31. Healing of Injured Bones.—A fractured bone takes from three to six weeks, and sometimes longer, to become healed. A dislocated bone, after it is reduced, requires to be kept quiet until all pain and swelling have subsided. In either case, there always remains more or less stiffness, which sometimes does not disappear for months after the accident.

32. Care of Injured Persons.—It frequently happens that a bone is broken when the person is at a distance from his home, or from any place where he can be attended by a surgeon. In fractures of the lower limbs, he must be carried often for a long distance, and every one should know how to make him comfortable during transit. It must be remembered that the only object of any person who is not a surgeon, should be to keep the broken limb in such a position that there will be no motion of the fractured ends, so that the patient may suffer as little as possible, and the surgeon may find him as nearly as may be in the condition in which the in-

verely strained that some of the ligaments are torn apart. Very often only a few fibers are ruptured, but such injuries always cause great suffering, and recovery is very slow. This form of injury is called a *sprain*, and is most likely to occur in the wrist, ankle, or knee.

jury left him.* Therefore, he should be carried on a litter, the broken limb being packed about with soft materials in such a way as to keep it from rolling or jarring. The weight of the foot will often make the lower part of the leg swing from side to side, and in the case of a fractured thigh, the leg should be protected on each side from the hip down. Dislocations require the same care, excepting that a splint is not necessary.

33. Effect of Alcohol on Growing Persons.— Besides the danger connected with the use of alcoholic drinks, which is common to them with other narcotic poisons, alcohol retards the growth of young cells and prevents their proper development. Now, the bodies of all animals are made up largely of cells, as heretofore shown, and, the cells being the living part of the animal, it is especially important that they should not be injured or badly nourished while they are growing. So that alcohol, in all forms, is particularly injurious to young persons, as it retards their growth, and stunts both body and mind. This is the theory of Dr. Lionel S. Beale, a celebrated microscopist and thinker, and is quite generally accepted.

34. Effect of Tobacco on Growing Persons.— Tobacco does not usually produce any permanent changes in the different organs as alcohol does. It seems mainly to influence the functions of organs,

---

* It is a very good plan to bind the lower limbs together in such a case, above and below the injured part, so that the sound leg may serve as a splint to the broken one. A broken arm may be bound to an improvised splint (a cane, a stick of wood, a shingle), a folded handkerchief or other padding being used to fill up the hollows between the splint and the skin, and the broken limb being supported by a sling around the neck.

without causing their degeneration, and the dys-
pepsias and other results of the use of tobacco
generally disappear promptly when the cause is
removed. This may at least be said of adults, but
it is not so true of the young. Any prolonged disturbance of the normal nutritive processes of the
body in a growing person produces permanent results, and children and youth are liable to be stunt-
ed physically and mentally by the use of tobacco.
The reason for this is plain. Growth and development go on according to age. They can not be
interrupted for months and years and put off until
a later period. If the normal development is hindered between fourteen and eighteen, and then a
healthier mode of life is resumed, the course of
development will not be taken up and carried on
as if the youth were still fourteen, but the effect of
his foolishness will be in great measure permanent.
A part of what has been lost may be recovered if
growth has not yet ceased, but it can not all be
made up.

35. Effect of Opium on the Young.—The great
difference in the power of resistance between the
tissues of the young and growing person and the
adult is strikingly shown by the effect of opium
upon them respectively. The danger of giving it
to children is so great that medical books are full
of cautions against it, and, while one grain is the
medicinal dose for an adult and four grains are
the smallest quantity known to have been fatal to
one, infants have been killed by one twentieth of a
grain.

CHAPTER IV.

MUSCLES.

36. The Muscles.—The bony framework of the body is set in motion by a system of organs called *muscles*, which cover the skeleton almost entirely (Fig. A), and cause the different bones to move upon each other by means of their peculiar property of contractility, or the power of becoming longer or shorter under varying circumstances.

There are two kinds of muscles in the body, called *voluntary* and *involuntary*, which differ very much in their structure and functions. The voluntary muscles, as the name implies, are under the control of the will; while the involuntary muscles are not only beyond our control, but act as a rule without our knowledge or consciousness.

37. The Voluntary Muscles.—A *voluntary* muscle is a mass of reddish fibers, somewhat loosely joined together by connective tissue, and easily separated lengthwise.* The flesh of animals is composed of muscular tissue. Every voluntary muscle is united

---

* If the fibers of a piece of lean meat are carefully separated and closely scrutinized, it will be seen that they are connected with each other by a delicate tissue of fine white threads, interwoven like the fibers of a cobweb or of the most delicate lace-work. This is called *connective tissue*, and is found in almost all parts of the body, uniting the different elements that make up the various organs.

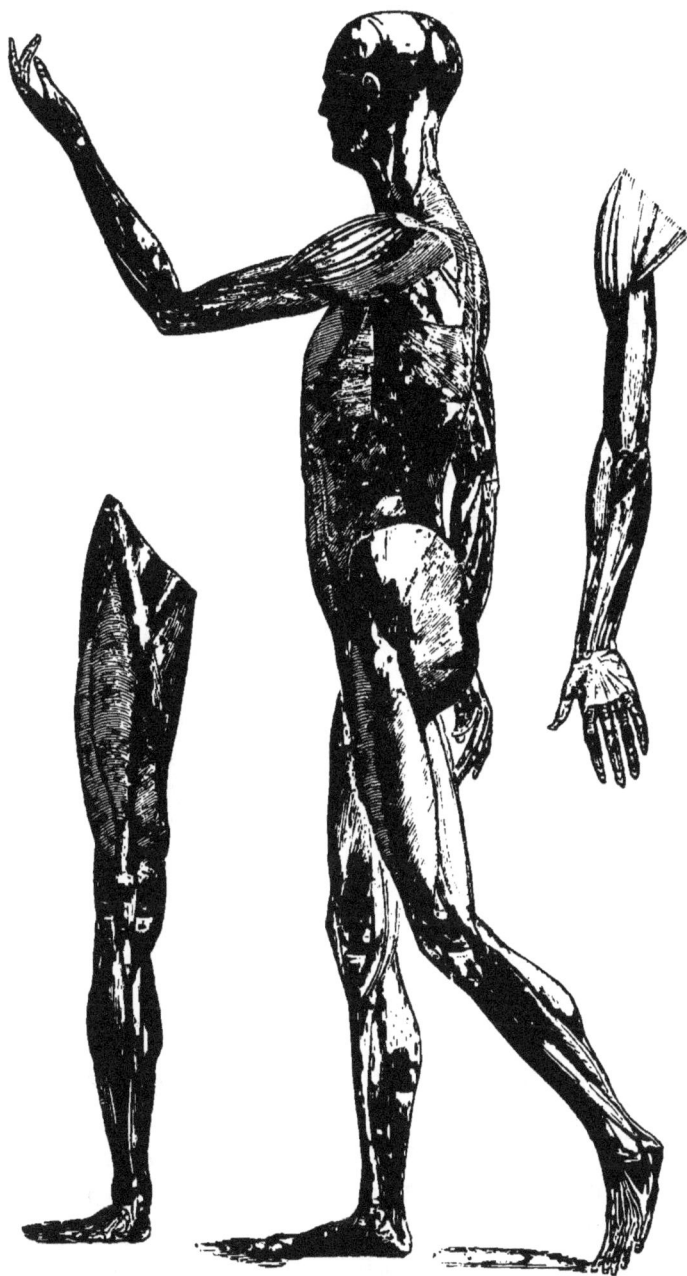

Fig. A.—The muscular system.

at each end to some fixed point in the body, and
there is always a joint or point of flexure between
its points of attachment. When the muscle contracts, therefore, the two ends are brought nearer
together, and motion is produced in the organ or
limb to which it is attached.

Every voluntary muscle can be divided into
small fibers, lying side by side, and these again into
fibrils still more minute. Each fibril under the
microscope presents an appearance of delicate lines
running at right angles to its length (Fig. 16).

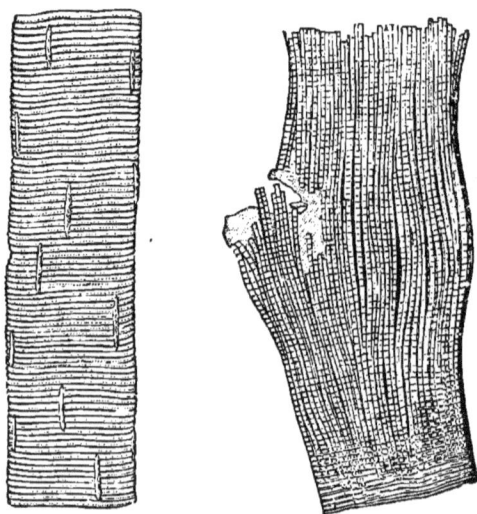

FIG. 16.—Voluntary muscular tissue.

These lines are called *striæ*, and the appearance is
called *striation*.

38. The Involuntary Muscles.—*Involuntary* muscles are made up of flattish bands of long, narrow
fibers, tapering at each end, somewhat of the shape
of an oat, but more slender. Each fiber has a nucleus in its middle, and they are all connected

together lengthwise, as the voluntary muscular fibers are (Fig. 17).

39. **Differences between the Voluntary and Involuntary Muscles.**—The *voluntary* muscles are all composed of the striated muscular fiber which allows of very rapid contraction, while the involuntary muscular fibers contract in a very peculiar manner. They do not begin to contract immediately, as soon as they are stimulated, but there is a short interval between the irritation and the response of the muscle. Then the contraction begins, and proceeds slowly and continuously up to a certain degree, when the fibers slowly relax, very much like the slow, crawling motion along the body of a worm or snake, when a wave seems to travel from one end to the other.

FIG. 17.—Involuntary muscular tissue.

Now there are some organs in the body, whose action must be rapid, from the nature of the office they perform, and still it would not do to have their motions depend upon the will. Such an organ is the heart. It must contract often and quickly in order to supply sufficient blood to the body, and yet, if its action depended upon our will, it would require all of our attention, to the exclusion of everything else. Accordingly, we find it composed of muscular fibers that are intermediate in structure between the voluntary and involuntary kinds. The involuntary muscular fiber is found, among other places, in the stomach and intestines, in the iris of the eye, and in the walls of the arteries.

40. Difference in Size of Muscles.—The largest muscle of the back, the *latissimus dorsi*, weighs several pounds; and one of the muscles of the leg, the *sartorius*, is two feet long; while the *stapedius*, one of the little muscles inside the ear, is only the sixth of an inch long, and weighs about a grain.* Between these extremes are many variations in size and shape.

41. The Tendons.—Muscles are connected with the bones by means of tendons. A *tendon* is made up of fibrous tissue, and is a white, glistening cord, of exceeding strength and toughness. At the ends, they gradually change their appearance, becoming muscle at one extremity and bone or periosteum at the other. There is no sharp line, where the muscle or bone can be distinguished from the tendon. Wherever the tendons would be likely to rise and form a line like the string of a bow during the contraction of a muscle, as at the wrist and the

---

* *Latis'simus dorsi*—i. e., the *broadest of the back*. This muscle is attached to the spine in the lumbar region and also to the lower ribs. The fibers come together so that the muscle has a triangular shape, and its small end is attached to the humerus. It is the chief muscle that comes in play when the body is raised from the ground by means of the arms.

The *sarto'rius* means the *tailor's muscle*. It is a long, ribbon-like muscle, which begins on the outside of the hip-bone and ends on the inside of the knee, crossing the thigh on the inner side. When it contracts, it raises the lower part of the leg, and turns it inward, thus crossing the legs, tailor-fashion—hence its name. It comes in play when one foot is placed on the opposite knee.

Stape'dius means the *stirrup-muscle*, so-called because it is attached to a small bone in the ear, which is shaped like a stirrup, and hence called *sta'pes* (Latin for stirrup).

ankle, for example, they are bound down by stout ligaments, through or under which they slide to and fro, the channels in which they move being lined with synovial membrane like the joints.

42. Force of Muscular Contraction. — When a muscle contracts (whether voluntary or involuntary), it becomes not only shorter and thicker, but harder, than before, and the *force* with which it contracts is enormous. To attain the compactness which we see in the body, the muscles of the limbs, for example, have to lie parallel with the length of the limb. Besides this, many of them are attached between the fulcrum and the weight, and very near the fulcrum. The biceps, for instance, which (with the brachialis anticus) bends the forearm upon the arm, is attached at one extremity to the shoulder-blade, and at the other to the forearm, just below the elbow, where its tendon can be felt. Thus there are two disadvantages under which it acts. In the first place, its point of action is only about one eighth as far from the joint or fulcrum as the hand is, and in the second place, when it begins to contract, it acts at a very acute angle—in fact, almost parallel with the bone (Fig. 18). As the arm becomes flexed,

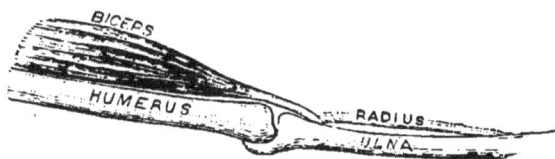

FIG. 18.—Disadvantageous action of the biceps muscle, illustrated.

the angle of action approaches more and more to a right angle, and the necessary effort becomes less

4

and less. And yet we not only flex the arm easily
enough at the elbow, but we do it with a considerable weight in the hand. It has been estimated that
the muscles of the arm, in flexing it at the elbow,
with a ten-pound weight in the hand, contract with
a force of at least two hundred pounds. And yet
this is a feat which a delicate woman or a child can
perform, and the force required is not to be compared with the power of an athlete.

43. Muscular Irritability. — Muscular tissue will
contract under any kind of irritation. In the living
body, the stimulus always comes from the nerves,
but the muscle itself has a form of *irritability*, which
lasts for a considerable time after death. When an
ox is killed, and has been prepared for the market,
the muscles may often be seen twitching and quiver-
ing in the beef for half an hour, and the muscles of
an amputated arm may also be seen to contract for
some minutes merely under the irritation of the cold
air. In cold-blooded animals, this irritability persists
for a long time. If the heart of a frog be entirely
removed from the body, it will continue to beat for
several minutes, and, when it has finally ceased, it
will start again on being pricked with a needle.
This experiment may be repeated several times before the muscular irritability finally vanishes.*

44. The Muscular Sense.—When a muscle contracts, the degree of contraction is perceived or felt

---

* There are reasons for believing that the continued beating of the
heart of a cold-blooded animal for hours after it has been removed
from the body may be due to the presence of microscopic nervous
ganglia in the substance of the muscle. This supposition, however,
does not affect the usefulness of the frog's heart as an illustration of
the fact that parts of animals continue to live after separation from the
main body.

by the brain. For example, any one is conscious whether his thumb is bent inward toward the hand, or outward toward the wrist, entirely apart from the use of the sight. The precise manner in which this sensation is conveyed to the brain is still a subject of conjecture. Although apparently so simple, it brings up questions of great intricacy and difficulty, which can not be considered here. But this sense, whatever its manner of operation, is called the *muscular sense.* It is one of the chief means we have of determining the weight or the hardness and softness of bodies, as we judge of these qualities mainly by the resistance our muscles meet with when handling the bodies. But, more than all, the muscular sense is necessary in keeping the body upright. The size of the feet is so small, compared with the height of the body, that early in life it is a matter of extreme difficulty for us to keep our balance. To stand and walk is one of the first and one of the hardest things we have to learn. It requires a constant contraction of the muscles, now one set and now another, in order to keep from falling.

45. Use of the Muscular Sense in Standing.— Ordinarily we are assisted in standing upright by our sight. This fact, together with the muscular effort required to stand still, may both be made very evident in the following manner: if a person stands with the feet close together, he will perhaps feel a slight swaying of the body, which has to be counteracted by muscular contraction. Perhaps no such swaying will be perceptible to him. But now, still keeping the feet close together, let him shut the eyes, when the swaying of the body will become much greater than before, and the constant muscular contractions, now here, now there, will be so plainly felt as to be disagreeable. In certain diseases, this muscular sense in the legs is lost, and then the person can stand with the eyes open, but if the eyes be closed he instantly totters and falls, for he then has nothing to guide him as to his vertical position.

46. Waste during Muscular Contraction.—The cause of muscular contraction is an unsolved problem. There is nothing in the chemical composition or the physical structure of the muscle which would lead us to expect to see it contract when irritated, if we knew nothing more about it. All we can say is, that it depends upon the composition of the muscular substance, and we know, also, that every contraction is accompanied by a loss of or change of material. In this way, our muscles are being continually used up, and if they were not constantly supplied with fresh nourishment by the blood, they would soon wear out and die. But the minute muscular fibers (or the cells composing them) not only perform their special function of contraction, but are able to choose and take up out of the blood their own proper food and appropriate it.

47. Muscular Overwork.—If a muscle is hard pressed and exercised too much, so that the waste of material is greater than the supply, and it wears away faster than it is repaired, it falls into the condition which we call *fatigue*, and it is only with great effort that we can make it work. If it be still further imposed on, without opportunity to recuperate, it soon gives out entirely, and can not be made to contract with vigor under any stimulus our brain can send to it. Such extreme fatigue is dangerous, because there is always the chance that the muscular fibers may become so completely wasted that even their power of nourishing themselves may be impaired, and the recovery of their natural condition may be very slow and imperfect, or, in rare cases, impossible.

48. Muscular Inactivity.—On the other hand, if a muscle is not exercised at all, its power of nourishing itself is interfered with almost as much as if it is exercised too much. It is found that unused muscles gradually waste away, growing smaller and smaller, and becoming soft and flabby, and finally, if they are not used for a very long time, it can be seen by the microscope that the muscular fibers disappear altogether, or are filled with little particles of fat, which take the place of some of the muscular substance, and so make it very weak and useless. Such inactivity of the muscles may occur in cases of paralysis, and the physician is then careful to stimulate them with electricity, in order to keep them, as nearly as possible, in a sound condition. The electrical current, in such cases, takes the place of the nervous stimulus, which naturally causes muscular contraction. The muscles of a broken limb, also, which have necessarily been idle while the bone was mending, are always very feeble for some time after the limb comes in use again.*

*Curvature of the spine, which is more frequent among girls than among boys, is often directly attributable to lack of exercise. The muscles of the back become weak, and, as some exercise of the muscles of the right side can not be avoided, so long as the girl performs any duties whatever, the difference in strength between the muscles of the right side and those of the left side becomes greater than is natural. The result of this is that the stronger muscles overpower the others and pull the spine over toward the right side, greatly distorting the figure. In left-handed persons the curvature is toward the left side.

49. Exercise.—It is necessary, therefore, that the muscles should be sufficiently exercised, and not too much. The kind of exercise is not of so much importance. No better form of exercise can be devised than the various out-door sports that boys are so fond of. It is much better that exercise should be a pleasure than a duty. For this reason, the ordinary exercises of the gymnasium do not compare in value, as health-giving ones, with rowing, skating, running, riding, wrestling, swimming, and the various out-door games.* It is really of no 'advantage, in our ordinary modern life, that the upper arm should, by judiciously and ingeniously planned exercise, grow to be an inch larger than it was a year before, and to the ordinary youth the duties of a gymnasium are irksome to the last degree.† There is no evidence that athletes, whose bodies are knobbed with unsightly bunches of muscle, are any healthier or any happier, or live any longer, or do any more good in the world, than the less muscular

---

* These remarks apply to girls as well as to boys. Out-door exercise of an agreeable kind is as necessary for the health of one as of the other. The hot-house plant is never strong, and the tom-boy grows to be the most healthy and vigorous woman, both mentally and physically.

† It is not to be understood that the gymnasium is here altogether condemned. It is of great use in its proper sphere. But the boy's idea of a gymnasium is that it is a place to get strong, rather than healthy. The surroundings and examples are such as to encourage straining for effect, lifting heavy weights in emulation, and the like acts, which may injure a boy permanently. When gymnastic exercises are performed under a competent instructor, with proper ends in view, and an intelligent use of means to those ends, the matter is altogether different. But, as mentioned in the text, gymnastic exercises, excepting for the purpose of remedying particular defects, training special muscles for a particular purpose, or treating actual disease, can not be compared in efficiency with out-door sports.

person who confines himself to simple food, who insists upon pure air, and exercises moderately and for his own pleasure in the way that suits him best.*

50. Danger of Exhaustion.—But, while muscular exercise is necessary to continued good health, it should never be carried to the point of exhaustion. This is dangerous, not only, as previously indicated, because the nutrition of the muscle may thus be interfered with, but because, when the point of simple fatigue is passed, and exhaustion supervenes, the nervous system has become implicated and is getting worn out. This danger will be better understood when that part of our bodies is described hereafter. It is enough, for the present, to remember that a person is not harmed by being tired, but that it *always harms one to be exhausted.*

51. Rest.—When a muscle is fatigued, it recovers very fast if allowed to rest. For this reason it is much less fatiguing to walk an hour than to

---

* The muscular strength which is developed by gymnastic training vanishes when the training ceases. It is often noticed by those who practice much in gymnasiums that constant practice is necessary to retain what increase of muscular power they have acquired. There seems to be a normal condition of the muscular system in each individual, to which he reverts when special training is abandoned. The strong men are not made so by training; they are born with a tendency to a preponderance of the muscular organs. Marvelous stories are told of men of this class. It is said of Frederick Augustus of Saxony, King of Poland (1670-1733), commonly called Augustus the Strong, that on one occasion, wishing to present a bouquet to a lady, and seeing nothing to wrap it in, he took a silver plate from the table and folded it around the stems with the greatest ease. In Dresden is exhibited a horseshoe, or the halves of it, which he is said to have broken with one hand. Similar stories are told of Baron Trenck (1711-1747); and of Milo, of Crotona, a famous athlete (520 B. C.), it is said that he once carried a live ox on his shoulders around the stadium, then killed it with a blow of his fist, and afterward ate the whole of it in a single day.

stand still an hour.   In the former case the muscles
constantly have short intervals of rest, while in the
latter they are not able to rest at all, but are continually in a state of contraction.   If we are obliged
to stand for a long time, therefore, we almost in-
stinctively change our position frequently, stand on
one leg and then on the other, or find some place
to lean against, in order to give the muscles the
rest they need.

*Out-door sports, then, are more healthful than gym-
nastic exercises.*

*Exercise may be pushed to the point of fatigue with-
out injury, but never to the point of exhaustion.*

**52. General Effects of Alcohol upon the Struct-
ures of the Body.**—When alcohol is introduced
into the system it circulates in the blood, and,
coming directly in contact with the different organs, irritates the delicate structures of which they
are composed; and, if used habitually, tends to produce gradual permanent changes in them, which
can mostly be brought under two heads, viz., an
increase of connective tissue or a deposit of fat.
Both of these tissues in the healthy body are chiefly
useful on account of their physical properties, the
connective tissue serving as a closely woven web or
frame-work to support the softer structures (cells,
etc.) and retain them in position, while the fat
makes up the padding of the body, and also forms
a reserve fund upon which the working tissues
draw for sustenance when the amount of food taken
is insufficient.   When these tissues are deposited in
abnormal quantities, however, they become great
agents of harm.   They encroach upon the neighboring cells, squeeze them out of shape, interfere

with their nutrition and the proper performance of their functions, and so diminish secretion and excretion, press upon and obstruct, or even obliterate, blood-vessels, irritate nerves, and in many other ways throw the delicate mechanism of the body out of gear. These alterations will be more particularly mentioned hereafter.

53. Effect of Alcohol upon the Muscles.—Fatty degeneration of the muscle is a common result of the use of alcoholic drinks, especially beer.

The microscope shows a striking difference between the firm, elastic structure of a healthy man's muscle and the pale, flabby, inelastic fibers of a heavy, inactive beer-drinker. Such a person may appear very strong and healthy, but his strength quickly deserts him upon any unusual exertion. It is found, also, that even moderate drinkers are more likely to be attacked by epidemic diseases, that they do not bear surgical operations so well, that they suffer more from exposure of any kind, and that they are apt to succumb to diseases from which the abstinent generally recover.

54. Alcohol diminishes the Power of Endurance. —It has been amply shown by Arctic exploration and by military campaigns in India and Africa, that those who use no alcohol endure privation, fatiguing labor, and extremes of temperature much better than those who take daily rations of grog. The common opinion that alcoholic liquors ward off the cold and temper the heat arises from the fact that the bodily sensations are dulled by the narcotic; the drinker, in other words, is partially anæsthetized, so that, although he feels cold and heat in a less degree, he is really less able to resist them.

PART III

ORGANS OF REPAIR.

CHAPTER I.

FOOD.

55. Necessity of Food.—We all know that, as long as we are living beings, we tend constantly to lose weight. In our excretions, our breath, the perspiration, the tears, the saliva, we lose altogether several pounds a day. All of this matter is so much gone, and if it be not replaced the body dies. It can not be too clearly impressed upon the mind that this waste or loss of material is continuous and inevitable. The processes of muscular contraction, of secretion, even of thought, produce substances which are taken up by the blood to be put out of the body. These substances are, many of them, very poisonous, and if they can not be expelled from the body they kill it.* They are not the result of disease ; they are the constant product of living processes in a healthy body. So we see

---

* Thus certain waste matters are excreted from the body through the kidneys. Some of them are very poisonous, and, when the kidneys are diseased and are no longer able to discharge them all from the body, they accumulate in the blood and finally cause death. So it is with the bile and with certain matters which pass away in the breath at every respiration.

that there must be a continuous expulsion of such
matters, and, of course, what each part of the body
has lost by such a process must be replaced with
fresh material.

56. Living without Food impossible.—If this fact
be clearly understood, it will be easy to see that the
numerous stories about persons who live without
eating are false. If such persons live, their hearts
must beat, their brains must think, their lungs must
move in breathing, and all of these things cause inevitably a waste of material. How absurd, then, to
gravely talk of a person who has not taken any
food or drink for six months, and still has not lost
weight, but remains plump and healthy! It is just
as absurd as it would be to say that such or such a
person had a limb amputated day after day, and yet
after each operation weighed as much as before.
These cases are all cheats, for if there is waste going
on, which is not made good, the body must decrease
in weight. If there is no waste, there is no life, no
thought, no heart-beats, no respiration, no movement of any kind. These facts of the generation of
force by food and of constant loss and gain are the
chief foundation-stones of all correct knowledge of
physiology, and can not be too firmly fixed in the
mind.

57. Classification of Food.—In order to supply
the waste in our bodies we need a great variety of
food ; and, indeed, the procuring and preparing of
food occupy a large portion of the lives of most people. The food we use is usually classified as *nitrogenous* and *non-nitrogenous*, or *carbonaceous*. But, besides
these two great divisions, which include all our animal and vegetable food, there are some substances

which are neither animal nor vegetable, and yet are quite as necessary to our health as any other portion of what we eat. The most important of these are water and salt.

58. Water.—*Water* is present in a greater or less quantity in every part of the body, and, as it is rapidly expelled, it has to be frequently supplied. It constitutes between three fourths and two thirds of the entire weight of the body, and the amount required for an adult man daily is about three pints, in addition to that which forms a part of the solid food. The quantity used varies enormously, according to the waste. In a hot day in summer we need much more than in cold weather, and in damp days much less than on dry ones.

59. Salt.—*Salt*, also, is not only an agreeable condiment, but has important offices to perform in the body. It has been shown by experiments on animals that, if they are entirely deprived of salt, they decline very much in vigor, and every farmer knows how necessary it is to the health of his cattle and sheep.*

60. Other Inorganic Matters.—There are other

---

* Boussingault, a French chemist (born in 1802), reported in 1854 some experiments he had made in regard to the importance of salt to cattle. He took six bullocks, of about the same age and vigor, and fed them alike, excepting that to three of them he gave 500 grains of salt every day and to the others none. At the end of six months the hides of those that had had no salt were rough and dull in color, while those of the others were shining and smooth. At the end of a year the salt-fed bullocks were in perfect health, while the others were dull and stupid, and the hair upon their hides was rough and tangled, with bare patches here and there.

Wild animals, especially of the grazing kind, like deer and cattle, will travel long distances in search of salt, and seem to be as fond of it as children are of sugar.

inorganic matters which are essential to the growth and nutrition of the body, but which are naturally found in articles of food and are not taken separately. Such are the salts of *lime, soda, potash,* and *magnesia,* all of which form a part of our common fruits and vegetables. The most important of these is probably the *lime phosphate* which forms so great a part of the bones. The husk of grain contains a certain proportion of this salt, and in growing children, in whom the cartilaginous portions of the bones are becoming ossified, wheaten grits or Graham bread is a very welcome and advantageous article of diet. It has been affirmed that the large size of the inhabitants of Kentucky is due to the fact that they live in a limestone region, and the water they use is strongly impregnated with lime. So large a proportion of lime taken into the body, at a time when the bones are forming and growing and hardening, is said to make them longer and stronger than they would be otherwise.

61. Non-nitrogenous Foods. — The *non-nitrogenous,* or, as they are sometimes called, the *carbonaceous* foods, are *sugar, starch,* and *fat.* These substances are all composed of carbon, hydrogen, and oxygen, in varying proportions, the sugar and starch taken in our food being mostly of vegetable origin, while the fat may be either animal or vegetable.

62. Starch.—*Starch* forms a part of all grains and most vegetables, sago, tapioca, arrowroot, etc., being almost pure starch, which has been extracted from the plants in which it is found. Rice contains about 85 per cent of starch, wheat about 70 per cent, and the potato about 15 per cent. This latter amount seems very small, but most of the remainder of the 100 parts of the potato consist of water, and starch really forms the bulk of the solid matter.

It is a peculiarity of starch that it is very easily converted into sugar. This is actually accomplished in the human body, during the processes of masti- cation and digestion, as will be shown hereafter.

63. Sugar.—*Sugar* is taken in our food in various forms, for it has not always the same chemical com- position. It is always sweet, and is always easy to recognize as sugar, but varies in its proportions of carbon, hydrogen, and oxygen. Thus we find that cane-sugar, milk-sugar, and grape- or honey-sugar (often called glucose*), all differ from each other. Sugar is taken partly as an addition to the food for the sake of improving its flavor, and partly as a nat- ural constituent of vegetables and particularly of fruits, some of which contain an enormous propor- tion of it. Figs, for example, are more than half sugar, and hardly any fruit contains less than 10 per cent.

64. Fat.—*Fat* is found in almost all parts of the body, and particularly just underneath the skin, where it serves to give rounded outlines to the form, and also undoubtedly acts as an elastic cushion to protect the parts beneath from injury. During life, owing to the warmth of the body, the fat is fluid and transparent; † but after death, as the body cools, it

---

* Glucose is now produced artificially in enormous quantities by the use of sulphuric acid and corn. When anything containing starch is boiled with this acid, the starch is converted into glucose, which is the kind of sugar found in fruits. Cane-sugar can be changed into glucose in the same way, and as a matter of fact it is changed into glucose in the act of digestion, so that glucose must be looked upon as that form of sugar that it is natural for us to take in our food.

† If the fingers be held close together in front of a bright light, the

becomes solid. The fat which is found in the body is not all taken in with the food, but a certain amount of it is formed in the body itself, in a manner which is not yet understood. Certain articles of diet tend to increase the amount of fat in the body. This is notably the case with starch and sugar. In sugar-growing countries, as the Southern States, it is a matter of common observation that the field hands grow fat and sleek during the sugar-season, and lose their superabundant flesh when the season is over. Articles of food which contain much starch also increase the amount of fat. The famous Banting system of treating corpulence is based on this fact, and consists mainly in depriving the patient of starchy vegetables, grains, and sugar.*

65. Nitrogenous Foods.—The *nitrogenous* portion of our food is also both animal and vegetable, but chiefly animal. The principal substances of this class are *fi'brin, albu'men,* and *ca'sëin.* They all contain a considerable amount of nitrogen, in addition to carbon, hydrogen, and oxygen, and are generally called by physiologists *pro'tëid* substances, or the *protëids.*

Cascin is found in large proportion in milk, from which it is extracted to form cheese, and the two

rosy tinge of their borders shows that they are to a certain extent translucent. The fingers of a corpse, under similar conditions, are opaque.

* Mr. Banting, the court undertaker, was put under treatment for corpulence by Mr. William Harvey, a London surgeon. He was allowed to eat any meat except pork, any kind of fish except salmon or eels, any vegetables except potatoes or rice, any kind of poultry or game, dry toast, fresh fruit, and tea without milk or sugar. When he began this diet in August, 1862, he weighed two hundred and two pounds, and a year after, he had lost forty-six pounds, and reduced his girth twelve and a quarter inches.

others are found mostly in the animal fluids, and in muscular fiber.

There are also substances very much like the animal albumen and casein which are found in vegetables, but they present slight chemical differences, although they probably answer nearly the same purpose in nutrition. Peas and beans contain a considerable quantity of the vegetable casein.

66. Necessity of Variety of Food.—It is necessary, for the preservation of health, that our food should contain a sufficient amount of these different kinds of matter. We must have water; we must have salt and the lime compounds mentioned above; we must have starchy substances (much the same to the body as sugar) and fats,* and we must have a certain amount of nitrogenous food. If one of these be lacking, the body soon feels it, and, although the person may not know precisely why he feels bad, he will often recover from his temporary disorder by a mere change of diet. The lack of any particular ingredient in our food is often indicated to us by a longing for it. We feel a strong desire to eat particular things and no others, and such a desire may generally be taken as a safe indication that the body needs them.

67. Paramount Necessity of Water.—Of all articles used for food or drink, water, in some form or other, is the most indispensable. Men can live much longer on water without food than on food without water. The celebrated French physiologist, Magendie, found that dogs lived eight or ten

---

* This is said of a healthy person. Excessive production of fat, as in Mr. Banting's case, is to be regarded as a diseased condition, and so requires special diet.

days longer, when supplied with water alone, than when they were deprived of both food and water. The pangs of thirst have been felt in a slight degree by almost every one, and it is the experience of those who have suffered from deprivation of food and water, in deserts and shipwrecks, that the tortures of thirst are much harder to bear than those of hunger.

68. Daily Amount of Food.—It has been found by Dr. Dalton, by experiments upon himself, that an adult requires food in about the following proportions:

Meat.....	16 ounces.
Bread......................	19 "
Butter, or fat...............	$3\frac{1}{2}$ "
Water	52 "

or about two pounds and a half of solid food and about three pints of liquid food daily. This is about the least amount which will keep him in good health.

69. Cooking.—Man does not take his food in the natural state, like other animals, but prepares it by *cooking*. This process is of advantage in two ways: it softens the hard parts of the food, such as beans, potatoes, and the various grains, and the fibrous tissue of meat; and it also develops a pleasant flavor by the action of heat, which excites the flow of the fluids of the mouth and stomach, and thus aids digestion.

5

ORIGIN AND NATURE OF ALCOHOLIC DRINKS.

70. Fermentation.—For many ages men have prepared certain intoxicating beverages without knowing their real nature, which has only been understood within the last thirty years. It is now known that these beverages are the result of the same general law of Nature that causes our bread to mold, our preserves to sour, and makes other articles of food unfit to eat.

When living matter ceases to live, its component parts are gradually decomposed or dissociated from each other, and form new combinations. The tree decays, the fruit rots, animal matter putrefies, and thus the surface of the earth is kept clear from an accumulation of lifeless matter that would prevent new growth. "One grand phenomenon," says M. Pasteur, "presides over this vast work—the phenomenon of fermentation." The process by which this work is accomplished is believed to be due to the presence of small living bodies, or micro-organisms. In appropriating from the substances they work upon the elements necessary for their own support, these micro-organisms give rise to new compounds, many of which are poisonous. They pervade Nature almost everywhere; the air is full of them. Vigorous living matter they are powerless to affect; but on dead matter their work begins.

71. Nature of Fermentation.—There are many kinds of fermentation, each having its own special micro-organism, and each micro-organism producing its own special result. That which takes place in proteids, giving rise to offensive odors, is usually called "putrefactive fermentation"; but the kind which results in the various intoxicating beverages just referred to is called "vinous fermentation."

The micro-organisms, or ferments of vinous fermentation, appear under the microscope as bead-like cells. Under the right conditions, in a liquid containing sugar, they grow and multiply rapidly by a process called budding. They act upon the sugar of the liquid, causing it to break up into two new substances, carbon dioxide, usually called carbonic-acid gas, and alcohol.

Neither of these substances is at all like the sugar from which it has been produced. It is a law of Nature that fermentation changes the character of a substance it works upon.* The carbon dioxide is a gas which passes rapidly out of the liquor.†

* Vinegar, a substance entirely different from wine or cider, is procured from these liquors by a fermentation called "acetous fermentation." This is brought about by the action of micro-organisms called "mycoderma aceti," which belong to a different family from those of vinous fermentation. During acetous fermentation the alcohol disappears, and acetic acid, the sour principle of vinegar, takes its place. Vinegar contains no alcohol.

† Carbon dioxide is the gas which imparts the effervescent quality to all drinks. It is freely soluble in cold water, more so under pressure, and when the liquid is relieved of pressure, as when soda-water is drawn from a fountain, or as the liquid becomes warmer, the gas is freed and escapes from it in bubbles. When any liquid containing

The alcohol is a liquid, and does not pass out of the fermenting fluid. It is a poison, and imparts its poisonous quality to any liquid containing it. It primarily affects the brain, and is therefore classed as a cerebral, or narcotic poison. Before the nature of alcohol was well understood it was considered a food; but we now know that it does not repair the waste of tissues as vegetables, meats, fruits, and other articles of food do. It contains no nitrogen, which is an essential element in nearly every tissue of the body, and it can not therefore be considered nutriment.

72. The various alcoholic beverages are mostly manufactured from two kinds of vegetable products, viz., the malt liquors from grain, the wines from fruits, and the distilled liquors from either. The process differs somewhat according to the material used, because fruits naturally contain sugar, while grains do not, and the brewer or distiller has to procure the sugar first.

73. **Manufacture of Malt Liquors.**—To manufacture malt liquors, the grain (generally barley, but often wheat, rice, or corn) is soaked with water until it is soft and swollen, and then piled in heaps in a warm atmosphere until it begins to sprout. This process (which takes place naturally in the ground when the seed is planted) changes a portion of the gluten into a substance called diastase, and this, in some way not yet clearly understood,

carbon dioxide is discharged into a glass, the bubbles of gas rise constantly from the sides and bottom, because the outside of the glass is constantly absorbing heat, and as warm water does not retain it in solution, it is constantly escaping from the warmest part of the liquid and rising to the surface.

converts the starch, which constitutes the greater part of the grain into a kind of sugar (*maltose*).* This sugar would naturally be appropriated as nourishment by the growing plant, but before that can occur the sprouting grains are dried and charred in a hot kiln, forming what is known as *malt*. The heat kills the germ, but the sugar remains. The malt is then crushed and mixed with water in huge vats, where the sugar is dissolved out of it, as well as various other matters which impart color and flavor to the liquid. The liquid thus obtained, called the *wort*, is then drawn off and boiled with hops, both for the purpose of imparting a desired bitter flavor to the beer, and also of preventing germs from developing within it. This new decoction is then plied with yeast (consisting almost entirely of the micro-organisms of vinous fermentation), which begin at once to decompose the sugar of the liquid. When their work has proceeded far enough for the purpose of the brewer, the liquor is drawn off into closed casks, and kept in a cool place until used.

74. **Malt Beverages.**—The beverages produced in this way are ale, lager-beer, weiss-beer, porter, and stout. Ale, porter, and stout have the largest portion of alcohol, as the process of fermentation is allowed to go further before they are used as beverages. Porter and stout owe their dark color to the greater charring of the grain, producing caramel, or burned sugar, in the malt. These liquors contain from four to ten per cent of alcohol. Lager-beer, as sold in this country (really schenk-

---

* This process is known as "saccharine fermentation," but is not believed to be due to the action of an organized ferment, such as the micro-organisms of putrefactive and vinous fermentation.

bier), generally contains from two and a half to five per cent, and weiss-beer about two per cent.

75. The Alcoholic Appetite.—It must be remembered that in whatever quantity, or wherever alcohol is found, its nature is the same. It is not only a poison, but a narcotic poison. It belongs to the same class with opium, chloroform, ether, hydrate of choral, etc.; one great peculiarity of which is that they never leave the body through which they have once passed in quite the same condition in which they found it. The person who has once taken them is apt to feel a desire to take them again. And this desire is not like the ordinary appetite for food. It is not that their smell or taste is agreeable, for the reverse is often the case. It is the after-effect that is sought. The oftener this desire is gratified, the more imperious it becomes, until finally the man is no longer master of himself; he neglects his daily affairs, and takes no interest in anybody or anything but plans for a fresh supply of the poison. The alcohol found in beer and other light liquors, though present only in small quantities, possesses this cumulative attraction for itself, which tends, sooner or later, to lead to excess. The craving for alcohol, when indulged, becomes a disease both of body and mind. The only remedy for the disease is to remove the cause, by adhering strictly to total abstinence. Such abstinence is also the only absolute safeguard against forming this unnatural craving. In view of this easily formed and more easily aroused appetite, the custom of flavoring food of any kind with any form of alcohol is a dangerous one that should be avoided.

76. Home-Made Beer.—Root-beer, small-beer, and various other domestic beers are all made by adding sugar to a decoction of roots or herbs. When the liquid is "alive"—i. e., full of bubbles— it is bottled, and of course contains alcohol (generally from one to two per cent) as well as carbon dioxide. Since it is the nature of a little alcohol to create an appetite for more, these are not harmless drinks.

77. Wine.—Wines are made by crushing fruits containing sugar and allowing the juice to ferment. No addition of yeast is necessary in this case, because the "bloom of the fruit" itself contains micro-organisms that start the fermentation as soon as the juice is pressed out from the fruit.*

The germs of these microscopic organisms, rest-

* The micro-organisms which excite vinous fermentation have received the generic name of *Saccharomyces.* There are several species of these. That which constitutes the yeast of beer is called *Saccharomyces cerevisiæ.* One species found on the grape, and thus introduced into the fermenting vat, is *Saccharomyces ellipsoideus.* "The species most widely diffused in nature, being found on all garden fruits as soon as they are ripe, is *Saccharomyces apiculata.* It rarely or never occurs on fruit before it is ripe, and if it has shown itself on early-ripening fruits, such as currants, gooseberries, or cherries, it is wanting on others which ripen later—plums and grapes, for instance, as long as they are still green. In the interval between two periods of ripening of the fruit, even in winter, it is found capable of development in the soil beneath the plants whose ripe fruit it attacks, and very rarely in any other place. Distinct spores are not known in this species, only the vegetative cells produced by sprouting. . . . The actual life-history of the plant is therefore very simple, and it is easy to conceive how it finds its way with or from the ripe fruit by the aid of wind or rain to the ground and is carried back with dust from the ground to the fruits, and its living through the winter in the ground is not at all surprising; but it has still to be explained why it is so rarely or never found on green fruit or in some other place."—(De Bary's "Morphology and Biology of the Fungi, Mycetozoa, and Bacteria," p. 357.)

ing on the outside and stems of fruits, do not exist in the saccharine juices within the fruits themselves. Hence there is no alcohol in whole fruits, vegetables, or grains as they come to us from the hand of Nature. The sugar of the fruit juices, acted upon by these micro-organisms or ferments, is broken up into carbon dioxide and alcohol, and thereby the juice of healthful fruit is converted into the treacherous and poisonous liquor called wine. Curiously enough, the alcohol produced by the growth of these organisms is hostile to their own development, and, after it amounts to a certain percentage (about seventeen per cent) of the fluid in which it is produced, further fermentation is arrested. Therefore, if the grape-juice contains a great deal of sugar, the alcohol produced finally prevents further changes, and a sweet wine is the result, generally very strong with alcohol. If the amount of sugar is small, a dry wine, containing less alcohol, is the product.

Cider is properly an apple wine, and is so called by the Germans (apfelwein). As in beers and ales, the alcoholic strength, which is at first slight, increases with age, until " hard cider " contains from five to ten per cent. Cider is a drink that has led many to drunkenness.

78. Other Fermented Liquors.—Perry is pear wine. Wines may be and are made from almost any fruit, as currants, raspberries, blackberries, gooseberries, oranges, etc., for all fruits contain sugar.

Koumyss, or milk wine, is made among the Tartars by fermenting mare's milk, and in this country cow's milk. It owes its effervescence to the presence of carbon dioxide, and contains from one to three per cent of alcohol, according to age.

79. Effervescent Drinks.—An essential fact to be borne in mind is that whenever and wherever a liquid containing sugar begins to cast off bubbles it is fermenting ; the bubbles consist of carbon dioxide, and there will inevitably be alcohol remaining in the liquor. So that all foaming, effervescing, lively liquids contain alcohol, provided the effervescence is the result of fermentation. In soda-water (so called) and most artificial mineral waters, the carbon dioxide is manufactured separately and pumped into the liquid by force. In drugs, like Seidlitz powders, etc., and in mineral waters, it is produced by chemical decomposition, bearing no relation to fermentation and not depending upon the presence of any micro-organism, nor accompanied by the formation of alcohol. But it will be readily seen that all domestic wines and beers, even what is called small-beer or root-beer, as they depend for their palatability upon the development of carbon dioxide, must contain alcohol as well, and are therefore unsafe beverages.

80. The Tendency of " Light Liquors."—The characteristic tendency of alcohol to create an ever-recurring demand for itself is shown by the liking for stronger liquors that almost universally follows the use of fermented liquors—even the " lighter ones," that contain only a small per cent of alcohol.*

---

* The London " Lancet " of December 15, 1888, in an article entitled " Beer and Drunkenness," says : " Drunkenness is seizing with a terrible grip the working population of Belgium. Belgium is only surpassed by Bavaria in the consumption of beer, 240 litres a year

In wine-making countries, where the people have unlimited access to unadulterated wines, including both the light wines and those heavier with alcohol, intemperance prevails to an alarming extent. This, according to the testimony of reliable witnesses, is the case in France and Persia, typical wine-growing countries.*

81. Distilled Liquors.—The stronger alcoholic liquors are derived from the weaker ones by the process of distillation.

Alcohol, and many of the oils and ethers which impart flavor and odor to the various alcoholic liquors, are converted into vapor at a lower temperature than water. Therefore, if a dilute solution of these substances is heated, the first ingredients to be driven off as vapor are the alcohol and volatile oils, and if the vapor is cooled rapidly as it comes away, it will be condensed into the liquid

being credited to each inhabitant; while Russia and Denmark alone surpass Belgium in the consumption of spirit, the average in Belgium being 13 litres a year per inhabitant, or about 50 litres a year per adult. Dr. Petethan stated at Liege in 1886 that there were 100,000 persons in Belgium who drink half a litre of gin per day, and no less than 50,000 who drink a whole litre. . . . The outlook is bad for Belgium if the state can do nothing to check this degrading vice."

* The Rev. J. S. Cochran, long a resident missionary in Persia, says, "In the wine-making season the whole village of male adults will be habitually intoxicated for a month or six weeks."

Mr. Labaree, also a missionary in Persia, writes: "If I had any sentiments favorable to the use of wine when I left America, my observation during the seven years I have resided in this paradise of vineyards have convinced me that the principle of total abstinence is the only safeguard against the great social evils that flow from the practice of wine-drinking. . . . There is scarcely a community to be found where the blighting influences of intemperance are not seen in families distressed and ruined, property squandered, character destroyed, and lives lost."

form again. This well-known fact affords a cheap and easy method of separating alcohol from the liquids containing it. The separation in this crude manner is never complete, because, as the temperature of the mixture rises, some water also passes off in vapor and is condensed with the alcohol; but a liquor is obtained which contains some forty or fifty per cent of alcohol, instead of the five or ten per cent contained in the fluid before distillation.

The distiller, at about the time when the brewer or vintner bottles up his product or puts it in closed barrels, heats it in a closed vessel, from the top of which a pipe conveys the vapor that arises through a vessel of water kept constantly cold. This pipe is generally made in a spiral or corkscrew form, so as to expose more surface to the cooling effect of the water, and for this reason has been called "*the worm of the still.*" The condensed liquid falls into a vessel prepared to receive it, and constitutes what is called distilled spirits or liquors.

82. Distilled Liquors as Beverages.—The most common of these liquors are brandy, rum, whisky, and gin. Brandy is distilled from wine, or the refuse grapes from which wine has been made. Rum is distilled from fermented molasses and water, or the fermented juice of the sugar-cane, or the refuse left from the manufacture of sugar. Whisky is made from malt, by fermenting it with yeast as the brewer does, and then distilling the alcoholic product. Gin differs from other liquors in having been twice distilled. First raw whisky is made, generally from barley, and then it is redistilled with juniper berries, which impart their distinctive flavor to the product.

Many other strong alcoholic liquors are produced by the distillation of fruit juices, as blackberry and peach brandy, and apple-jack or cider brandy, distilled from cider.

All these distilled liquors contain from forty to sixty per cent of alcohol, and are therefore correspondingly destructive to health, character, and life. Their distinctive flavor and odor are due to various volatile oils and ethers which pass over with the alcohol as vapor, and are condensed with it, and the color is usually imparted by the wood of the cask in which the liquor is drawn off or by coloring-matters, caramel, etc., intentionally added to it.

83. Danger of Alcoholic Beverages. — In all these beverages the alcohol is the dangerous constituent, and during all the manipulations to which they may be subjected it remains a poison unchanged. Its various injurious effects upon the human system are not due to the weakness of the drinker, but to the nature of the drink.

CHAPTER III.

84. The Digestive Apparatus.—The food we eat is mostly insoluble, and in an unfit condition to be used for the nourishment of the body. Even so nutritious a substance as albumen can not be used without undergoing some change, and if pure fluid albumen be injected directly into the blood, it will be thrown out of the body by the kidneys unaltered. To prepare the various foods for use in the body, we are provided with a complicated series of organs, called the *digestive apparatus*, in which the food is ground fine and mingled with various juices until it is reduced to a fluid mass, which can be taken up by the blood and carried to all parts of the body in a condition fit for their nutrition.

85. Processes to which Food is subjected in the Body.—The process of preparing food for our nourishment may be conveniently divided into five stages. The *first* of these is *mastication*, which takes place in the mouth, and is a voluntary act. The *second* is *swallowing*, or the act of passing food on from the mouth to the stomach, the beginning of this act being voluntary, and the greater part of it involuntary. The *third* is *stomach digestion*, which is involuntary; the *fourth*, *intestinal digestion*, which

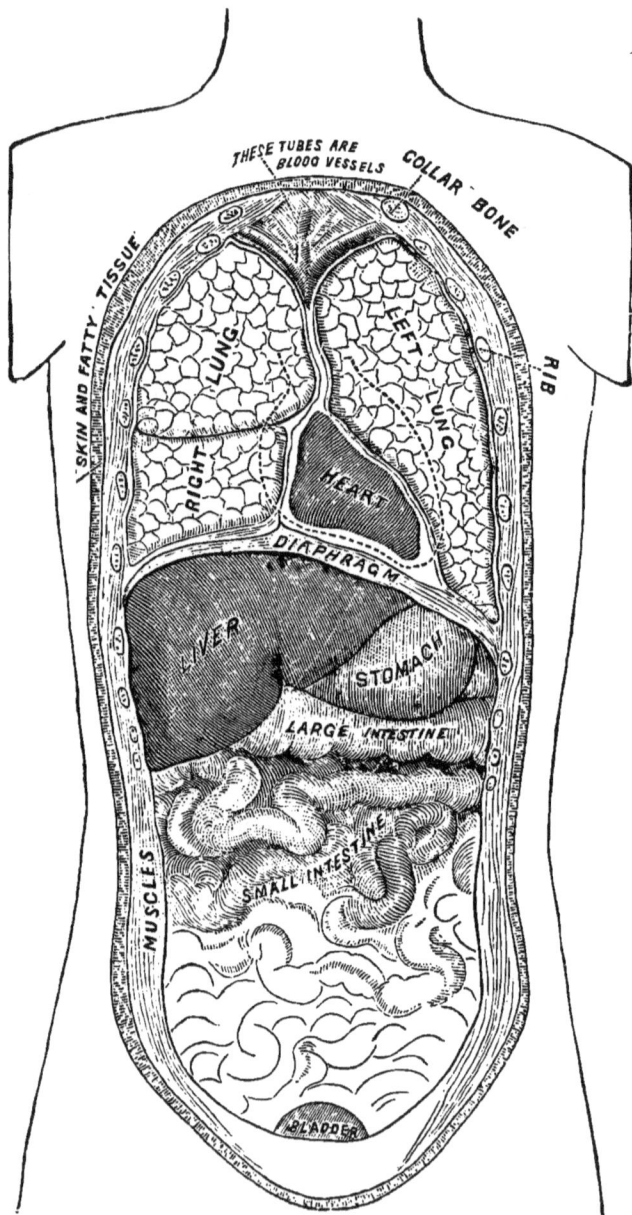

FIG. 19.—Front view of the organs in their natural relations. The heart is partly covered by the lungs, but its true outline is indicated by a dotted line. Only ten ribs are shown on each side, the eleventh and twelfth (the floating ribs) being too short to be included in the section.

is involuntary; and the *fifth* is the process of *absorption*, which is also a process of selection, by which certain portions of the prepared mass are taken up into the circulation for food, and the rest left in the intestines as waste material; this whole process is also beyond the control of our will; all waste matters are then expelled from the body.

86. Only One Voluntary Process.—All of these processes must be properly conducted, in order to maintain the body in a healthy condition. They are all important, and, if one be neglected or carried on in a disordered and unnatural manner, the others will all be affected, by reason of their close connection with and dependence on each other. Only one of them is directly under the control of the will, and every one can do more toward preventing dyspepsia and other disorders of the digestive organs, by paying some attention to the proper and complete mastication of his food than in almost any other way. If we examine the *mouth*, with reference to its uses in mastication, we find it prepared to perform *three* great and important functions.

87. Use of the Taste and Smell.—In the *first* place, it is provided with an organ of *taste*, to assist us in selecting our food. In this it is aided by the proximity of the nose, so that we have the additional advantage of the sense of *smell*. And the nose is not only so situated as to aid us in judging of food before it enters the mouth, but it is connected with the throat behind, so that odors are detected from substances already in the mouth.

88. The Teeth.—In the *second* place, the mouth is provided with organs for grinding and crushing the hard parts of the food, and reducing them to a soft mass, fit to be acted upon by the fluids in the stomach and intestines. The organs directly of use in this operation are the *teeth* (Fig. 20), but essential aid is afforded by the muscles of the *cheeks* and the *tongue.* The rows of teeth are narrow, and, except for the action of these muscles, the food could not be kept between them. Indeed, it has been found that in cases of paralysis, when the muscles of the cheek are unable to contract, while the tongue still retains its power, the food gets pushed out between the cheek and the teeth, and accumulates there. The lower jaw is moved by some of the most powerfully acting muscles in the body. The chief one of all is the *mas'-seter,* which is attached above to the ridge of bone running backward from the lower border of the eye toward the ear, and below to the horizontal portion of the jaw. The muscle is nearly square in shape, and, as is easily seen, acts at a great mechanical advantage. As a matter of interest connected with this muscle, it may be stated that it is the muscle generally referred to for proof that muscular contraction is accompanied by a sound. If the lower jaw be firmly closed, and the teeth powerfully pressed together so that the muscles of mastication are strongly contracted, a low, rumbling sound will be heard, which can not be explained in any other way than as caused by the muscular contraction.

FIG. 20.—Section of a tooth. The black portion is the cavity occupied by the nerve and blood-vessels.

89. **The Saliva.**—In the *third* place, the food, while undergoing mastication, is mixed with cer-

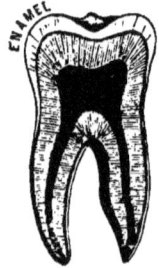

tain fluids, called collectively the *saliva*. They are
the product of *three* sets of glands, each of which
is double—i. e., there are three glands on each side
of the mouth (Fig. 21), and the secretion of each

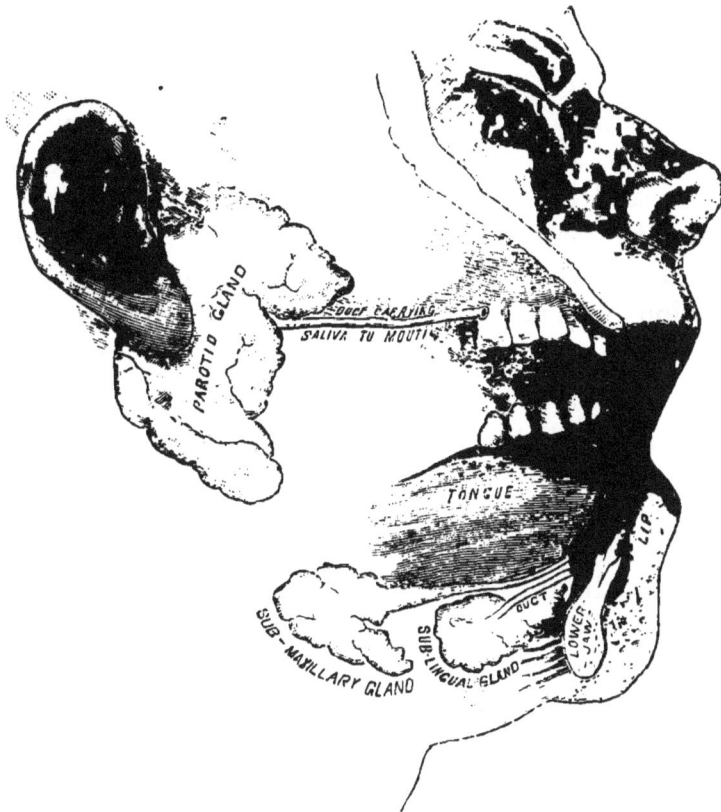

FIG. 21.—The salivary glands of the right side.

pair is peculiar to itself. The largest of these are
the *parot'id glands*, which are situated just in front
of the lower border of the ears, and are the glands
which become swollen and cause the distortion of
the face in the disease known as the mumps. The
fluid secreted by these glands is very thin and
watery, and constitutes the greater part of the sa-

liva. The other glands * are situated just inside the lower border of the jaw and beneath the tongue. Their secretion is much thicker and more glutinous than that of the parotid glands. Besides these fluids, there is a small amount secreted by the mucous membrane lining the mouth, and all these mingled fluids constitute the saliva.

The saliva is secreted to some extent at all times, and keeps the lining membrane of the mouth moist and soft, but it is a familiar fact that its amount is greatly increased at certain times. Thus, we say, at the sight, or even sometimes at the suggestion, of an appetizing meal, the " mouth waters." †　The fact of its excessive secretion at such times shows that it has a part to perform in the process of mastication and digestion.

90. Properties and Use of the Saliva.—Experiments have shown that saliva possesses the property of converting starch into sugar ; but, as this is also done by the digestive fluids, it is not considered to be a very important function, and the chief use of the saliva undoubtedly is, to make the processes of mastication and swallowing of food easier. If food were taken dry, and there were no means at hand of moistening it, mastication would be very difficult and tiresome, and swallowing almost impossible. Bernard ‡ found, by experiments upon a horse, that

* Submaxillary and sublingual.

† In physiological lectures before medical students, it is not uncommon to illustrate this fact in a curious way. A slender tube is introduced into the opening by which the parotid saliva is discharged into the mouth, and, according to the condition of the person operated on, there will either be no flow of saliva or it will come out drop by drop. But now let food be brought in, and the moment it is seen the saliva begins to run from the tube in a plentiful stream.

‡ Claude Bernard, a famous French physiologist (1813-1878). He

when an operation had been performed, which prevented the parotid saliva from entering the mouth,
the animal could masticate and swallow (the latter
process being accomplished with great difficulty)
only three quarters as much oats in twenty-five
minutes as he had previously eaten in nine. It is
probable that the parotid saliva, which is almost
like water, assists mainly in the mastication of the
food; while the other secretions, which are thicker
and more slippery, coat the outside of the mass, and
render it easier to swallow.

The total amount of saliva secreted by a healthy
adult in twenty-four hours has been calculated, after
numerous experiments, to be nearly *three pounds.*

NOTE.—It is important for the health of the individual that the
teeth should be kept in good condition, in order that mastication may
be thorough and complete. Particles of food which stick between the
teeth, if they are allowed to remain, putrefy and impart an offensive
odor to the breath. The acids which are developed during the putrefaction of such matters are also injurious to the teeth and tend to hasten
their decay. There are also certain substances deposited from the
saliva around the necks of the teeth, called "tartar." If this is not
removed, the gums are bruised against it, and finally recede from the
teeth, leaving a part of the fang bare, and thus exposing to all sorts of
injurious influences a part of the tooth which was never intended to
be so exposed.

The teeth should therefore be frequently cleaned, at least twice a
day, with water and a *soft* brush (a stiff brush injures the gums), tooth-
picks being used when necessary. Avoid fancy tooth-powders and
washes, for they often contain injurious acids or gritty substances. To
polish the teeth, use powdered orris-root and chalk, which can be
bought of any druggist. Never crack nuts with the teeth, and, on the
slightest appearance of decay, consult a good dentist.

was Professor of Physiology in the College of France from 1855 until
his death. Especially distinguished for his discovery of the formation
of sugar in the liver, and for his researches on the functions of the sympathetic nervous system. (See later.)

91. The Alimentary Canal.—The food, which is now ready to be operated upon by the digestive fluids, passes beyond the control of the person who swallows, and begins its travels in the long tube, called the *aliment'ary canal* (Fig. 22). This canal begins with the mouth (already described) and ends with the large intestine, and is nearly thirty feet in length. It does not properly assume the form of a tube until the beginning of the *œsoph'agus*, or gullet, and at one point, namely, at the lower extremity of the œsophagus, there is a considerable enlargement, which has the appearance of a bag or pouch, and is called the *stomach*. To understand the working of the alimentary canal, it is necessary to know something of its anatomy.

The two most important tissues in its structure are the mucous membrane which lines it throughout, and the muscles which surround it, and are imbedded in its walls.

92. Mucous Membrane.—*Mucous membrane* (Fig. 23) is the skin which lines the interior canals of the body. While the outside of the body is covered by a smooth, white, tough skin, we see that, at the openings leading to its interior, such as the mouth, nose, etc., the character of this covering suddenly changes, and it becomes a reddish or pinkish mem-

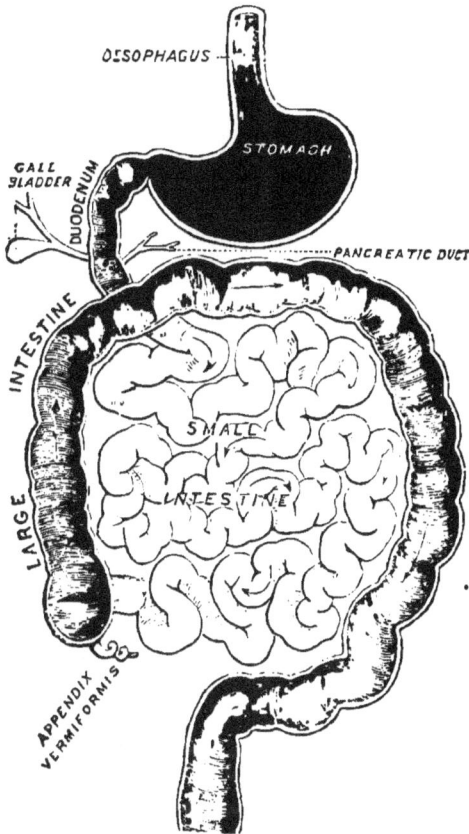

FIG. 22.—The alimentary canal.

brane very soft and delicate in texture, and continually moistened by its secretions. This is called mucous membrane, and in one form or another it lines all those internal parts of the body, which communicate with the external air.

It is made up largely of *fibrous tissue*, consisting of fine threads, interlacing with each other in every direction and densely woven. Its surface is covered with minute cells, called *epithe'lial cells.** At various points on the membrane are minute tubes or cavities, less than $\frac{1}{100}$ of an inch in diameter, of different shapes in different places, and in some situations so numerous that they lie almost in contact with each other. These minute tubes are closed at the bottom, but

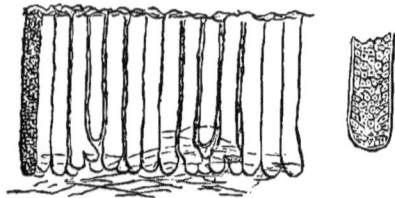

Fig. 23.—Structure of mucous membrane illustrated. At one side is a detached portion of a tube, or follicle, enlarged so as to show the epithelium more clearly.

open on the surface of the membrane. Small as they are, they are lined from top to bottom with epithelial cells, which really carry on the work of secretion. All around, among, and underneath these tubes are small blood-vessels, which nourish the membrane, and from which the little epithelial cells separate the materials which form the mucus.

93. Muscles of the Alimentary Canal.—The muscles which form a considerable part of the walls of the alimentary canal are of the involuntary or non-striated kind. The fibers run in various directions, some of them surrounding the œsophagus and the

---

* All free surfaces of the body, whether inside or outside, are covered with cells. In the interior of the body, the alimentary canal, the lungs etc., these cells are soft, and, so to speak, plump, and are called *epithelial cells*, or a mass of them taken together is called *epithelium*. On the external surface of the body they are dry, flat, hard, and horny, and are called *epider'mal cells*, or, in a mass, the *epidermis*. In both situations they are being constantly shed and renewed. The fresh ones are continually forming underneath, and, as they grow, take the place of the old ones on the surface, which are being constantly rubbed off in one way or another. All the secretions of mucous membranes contain these epithelial cells, and the slightest scraping of the skin dislodges epidermal cells.

intestines in a circle, so that when they contract they make the canal smaller; while others run lengthwise, and their contraction shortens the canal. When these two kinds of fibers, the circular and the longitudinal, contract together, they propel forward anything that comes within their grasp.

94. Serous Membrane.—Besides these parts of their structure, the stomach and intestines are covered on the outside by what is called a *serous membrane*, which is found lining all cavities inside the body that do not communicate with the air, excepting the joints. This kind of membrane is transparent, exceedingly fine and soft, and smooth like satin, and is constantly moistened with a slight amount of fluid. The use of serous membrane is to allow organs to move freely upon each other without friction. If it were not for some provision of this sort, the movements of the stomach and intestines during digestion would be painful, or at least disagreeable, while, as things now are, we are entirely unconscious of any movement at all.

95. Swallowing.—After mastication is completed the tongue passes the mass of food backward into the *phar'ynx* (or throat), whence it goes on into the œsophagus. The *œsophagus* (Fig. 24) is about nine inches long, and extends from the throat to the stomach, not just behind the breastbone, as many suppose, but just in front of the spine. The muscles of the upper portion are of the striated variety, but, nevertheless, their contraction is not voluntary. When anything has once passed to the back of the throat, it will be swallowed and sent into the stomach, in spite of our will.

96. The Stomach.—The *stomach* varies in size

in different persons, but on the average will contain about *three pints* of fluid in the adult. Its

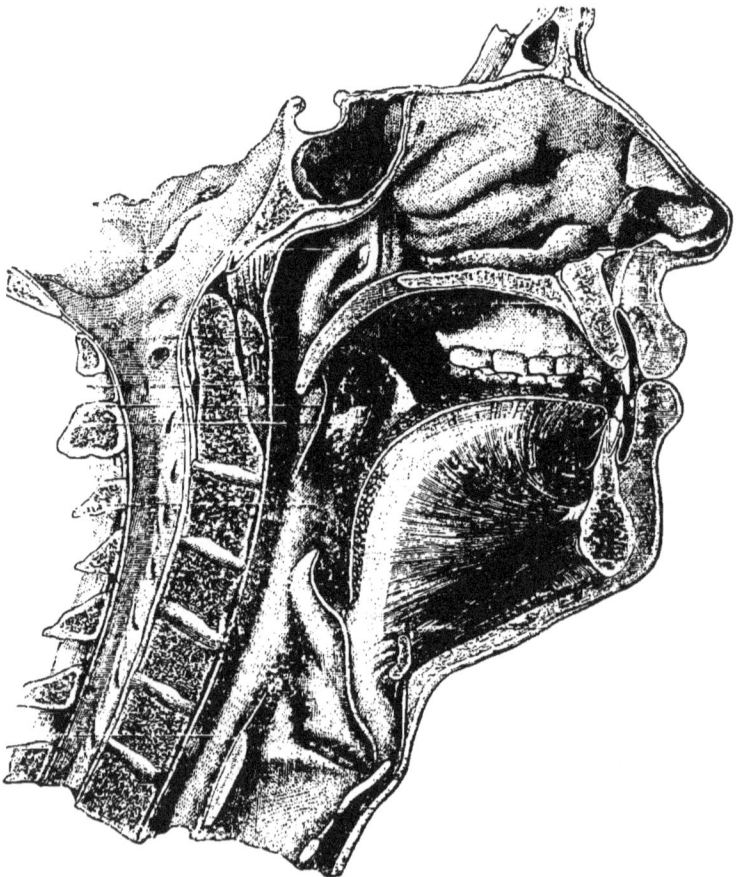

FIG. 24.—Vertical section of the head and neck. At the base of the tongue
is seen the epiglottis, and below this the larynx. Between the larynx
and the bodies of the vertebræ lies the œsophagus.

shape has often been compared to that of the air-bag of a bag-pipe, which it much resembles (Fig. 25). It has two openings, one at the lower extremity of the œsophagus, where food enters, and the other at the point where food passes out and the small intestine begins. These openings are both in the upper border of the organ, and only a short distance apart, the *pylo'rus*, or exit, being at the right

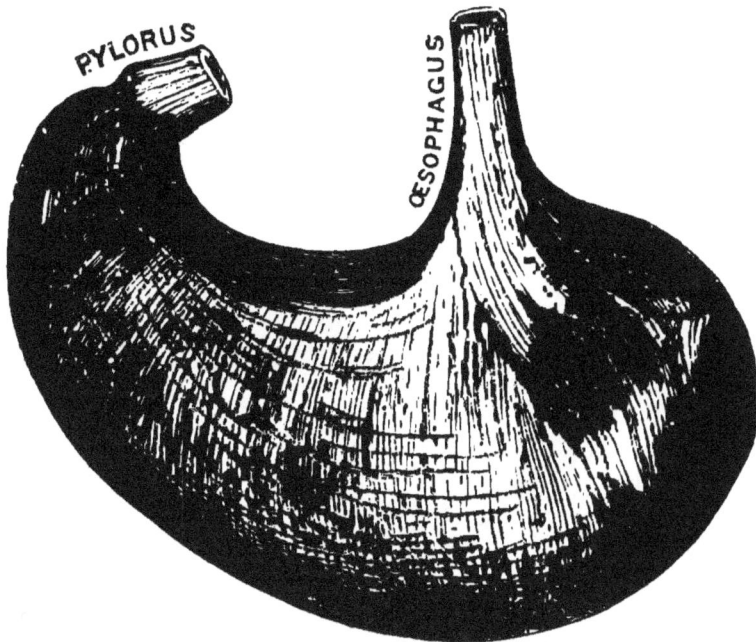

FIG. 25.—Outside of the stomach, front view, showing the muscular coat.

extremity, and the *car'diac* opening near the middle.* The stomach extends toward the left for about three inches beyond the cardiac orifice, and is larger in this part than in any other. This portion is called the *great pouch* of the organ (Fig. 26).

Each orifice is guarded by a powerful muscle, surrounding it in a circular form, which can contract so tightly as to prevent the passage even of a fluid. As a rule, these muscles prevent the passage of any substance backward through them, in opposition to the natural course of the food.

* *Pylorus*, a Greek word meaning the gate-keeper ; *cardiac*, from a Greek word meaning the heart, because it is very near that organ.

97. Stomach-Digestion. — It was formerly supposed that the whole process of digestion was

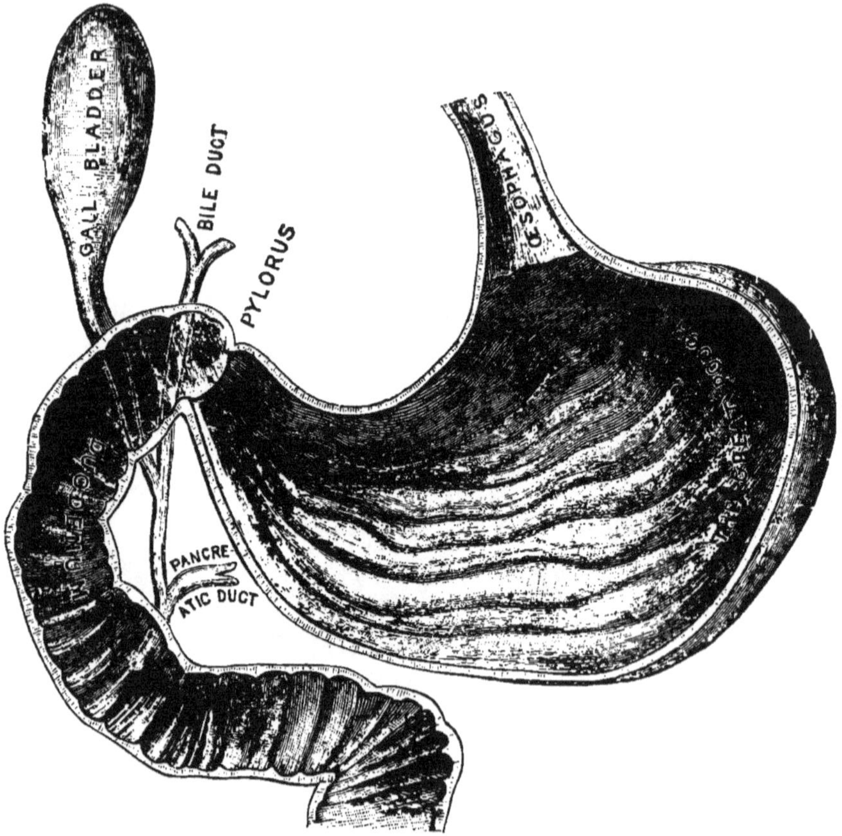

FIG. 26.—Inside of the stomach, front view, showing the folds (or rugae) of the mucous membrane.

performed in the stomach, but this is now known not to be the case. The nitrogenous portions of the food are the only ones that are digested in the stomach. The *oily* and *fatty*, as well as the *starchy*, portions are digested in the small intestines. Fluids are very rapidly absorbed by the stomach.

98. Dr. Beaumont and St. Martin.—There are so many difficulties connected with the investigation of the subject of digestion, that very little was really known about it until the year 1833, when a small book was published by Dr. Beaumont, of the United States Army, giving physiologists their first precise knowledge of what takes place in the human stomach. His observations were so well taken, that very little has been added since to what he discovered upon the particular subject of stomach digestion.

In the year 1822, Alexis St. Martin, a stout young French Canadian, in the employ of a fur-trading company, and about eighteen years of age, received a severe wound in the left side from the accidental discharge of a shot-gun at a distance of about three feet. He was attended by Dr. Beaumont, and, although his recovery was slow, his health was finally completely re-established, and he was still living in Vermont, the father of a numerous family, at a very recent date. In the situation of the wound, however, was left an opening into the stomach, about four fifths of an inch in diameter, closed by a flap or valve of mucous membrane on the inside. This valve could be pushed inward, but not outward; and thus, although the operation of digestion was not at all interfered with, the interior of the stomach could be thoroughly examined, and experiments performed with the greatest facility and accuracy. Dr. Beaumont kept the young man in his employ for several years, and made hundreds of observations upon him. These were published in his little book, and made both him and St. Martin immediately famous.

As Dr. Beaumont was the first, and for many years the only, person who ever saw the interior of the stomach in a living man, much of the following description will be taken from his volume.

99. Interior of the Stomach.—"The interior coat of the stomach," he says, "in its natural and healthy state, is of a light or pale-pink color, varying in its hues according to its full or empty state. It is of a soft or velvet-like appearance, and is constantly covered with a very thin, transparent, viscid mucus, lining the whole interior of the organ."

100. The Gastric Juice.—The changes which the food undergoes in the stomach are due to the action of the *gastric juice*, the appearance of which, with the manner of its secretion, is thus described:

"By applying aliments or other irritants to the internal coat of the stomach, and observing the effect through a magnifying-glass, innumerable minute lucid points can be seen arising from the mucous membrane, and protruding through the mucous coat; from which distils a pure, limpid, colorless, slightly viscid fluid. This fluid is invariably distinctly acid."

"The fluid so discharged is absorbed by the aliment in contact, or collects in small drops and trickles down the sides of the stomach to the more depending parts, and there mingles with the food or whatever else may be contained in the gastric cavity."

"The gastric juice never appears to be accumulated in the cavity of the stomach while fasting. When aliment is received, the juice is given out in exact proportion to its requirements for solution, *except when more food has been taken than is necessary for the wants of the system.*"

101. Composition and Amount of Gastric Juice.—

The *gastric juice* contains two important constituents, viz., hydrochloric acid and pepsin. If it be deprived of these, it will no longer exhibit its peculiar properties; while, if it retains them, as Dr. Beaumont first showed, it will digest food in a glass tube, outside the body, provided the tube and its contents be kept at a temperature of 100° Fahr., which is about the ordinary temperature inside the stomach.

The average amount of gastric juice secreted daily by an adult human being has been estimated at a little less than *fourteen pounds*, or about a *gallon and a half*.

102. **Movements. of the Stomach.** — But, besides the action of the gastric juice in stomach-digestion, a very important office is performed by the muscles which form a large part of the walls of the organ. During digestion, these muscles are continually contracting in a slow, regular order, producing movements of the contents of the stomach in a very peculiar manner, which, in health, never varies. Dr. Beaumont says that the ordinary course and direction of the revolutions of the food are first, after passing out of the œsophagus into the stomach, from right to left, thence down along the great curvature, from left to right, to the pylorus, whence it returns again along the upper border of the organ to the left extremity of it. Each of these journeys of the food around the organ occupies from one to three minutes, and they serve to mingle the gastric juice more thoroughly with the food. As soon as the process of digestion is gone so far as to bring portions of the food into a condition for absorption, it is found that every time the contents of

the stomach pass the pylorus the mass becomes diminished in amount, showing that a portion has
been squeezed or pressed through the opening into
the intestine.

" These peculiar motions and contractions continue until the stomach is perfectly empty, and not
a particle of food remains.   Then all becomes quiet
again."

Thus that part of the digestion of food which is
carried on in the stomach is accomplished by the
action of the gastric juice, and the changes produced by it are assisted, and the prepared food is
passed out of the stomach, by the constant contractions and churning motions of the organ just described.

This, then, is the ordinary healthy process of
stomach-digestion, when not in any way hindered
or interfered with.   Let us see what changes take
place in the appearance of the stomach and in its
functions when it is injuriously affected.

**103. Appearance of the Stomach during Indi-
gestion.**—" In a feverish condition, from whatever
cause—obstructed perspiration, undue excitement
by " alcoholic " liquors, overloading the stomach
with food—fear, anger, or whatever depresses or
disturbs the nervous system," the lining of the stomach " becomes somewhat red and dry, at other times
pale and moist, and loses its smooth and healthy
appearance; the secretions become vitiated, greatly
diminished, or entirely suppressed."

" There are sometimes found, on the internal
coat of the stomach, eruptions or deep-red pimples.
These are at first sharp-pointed and red, but frequently become filled with white purulent matter;

at other times, red patches, from half an inch to an inch and a half in circumference, are found on the internal coat. These appear to be the result of congestion in the minute blood-vessels of the stomach."

" These diseased appearances, when very slight, do not always affect essentially the gastric apparatus. When considerable, and particularly when there are corresponding symptoms of disease, as dryness of the mouth, thirst, furred tongue, etc., no gastric juice can be extracted. Drinks received are immediately absorbed or otherwise disposed of; none remaining in the stomach ten minutes after being swallowed. *Food, taken in this condition of the stomach, remains undigested for twenty-four or forty-eight hours or more.*"

" Whenever this morbid condition of the stomach occurs, with the usual accompanying symptoms of disease, there is generally a corresponding appearance of the tongue. When a healthy state of the stomach is restored, the tongue invariably becomes clean."

These are the observations of one who saw what he describes, and took careful notes of what he saw.

104. Time required for Stomach-Digestion.—The *time* required for digestion in the stomach varies very much according to the character of the food. Dr. Beaumont found that the time of stomach-digestion in St. Martin varied from one hour to about five and a half. Among meats, the soonest digested was boiled pig's feet, which took an hour, and the longest time was taken for roast pork, viz., five hours and a quarter; among vegetables, rice is digested in an hour, while boiled cabbage requires four. The average time required for an ordinary meal is probably about *three hours.*

105. Advantage of Thorough Mastication.—Dr. Beaumont found that, when a piece of meat or other food is attacked by the gastric juice, it is slowly dissolved from the outside. The juice is not soaked up and does not penetrate the interior of the mass, but gradually softens the exterior of it; and, as the outside portion becomes friable and dissolves, the piece grows smaller and smaller, the gastric juice in this way advancing little by little, until the whole mass is liquefied. From this it is evident that it will take longer to digest a large piece of meat than to digest the same amount after it has been divided into small pieces ; for this reason it is important to *masticate the food thoroughly before sending it into the stomach.*

106. Eating too little.—It is evident that it will not do to take too little food. Enough must be eaten to supply the needs of the system, and it must be of such a quality that it can be readily digested and appropriated.

107. Eating too much.—But, on the other hand, we *must not take too much food.* There seems to be some subtile relation between the amount of food required by the system and the amount of gastric juice furnished by the stomach. What is likely to be the result, then, if more food is taken into the stomach than can be acted on by the gastric juice? Let us consider. The temperature of the interior of the stomach is about 100° Fahr. This is just about the temperature at which fermentation and putrefaction (which is one kind of fermentation) are most active. Heat and moisture favor these processes. Both of these conditions exist in the stomach, but, under ordinary circumstances, the gastric juice prevents any other changes than those due to its own action. But, if more food is introduced than the gastric juice can dissolve, fermentation occurs, and offensive gases and irritating acids are produced. Then the symptoms of indigestion come on, there is constant belching of wind from the mouth, an uneasy sensation in the stomach, and, as soon as the undigested and fermenting mass passes out into the intestine, rumblings and colic set in, followed probably by a diarrhœa, which continues until the offending matters have been ejected from the body.

108. Eating between Meals.—Similar symptoms may be produced by *eating between meals.* When a sufficient meal has already been eaten, we should wait until it has been digested and the stomach has had a short period of rest before we give it any more work to do. This organ can not work incessantly any more than other parts of the body, and when it is ready for more food the sensation of hunger apprises us of the fact. If we load it with fresh food before the previous supply has been disposed of, there may not be enough gastric juice secreted to digest it. Then it ferments, or putrefies, and causes a fit of indigestion, as just described.

109. Hunger.—It is sometimes said that a person should rise from the table, after every meal, still hungry. This is not correct, and the reason is plain. Hunger is the natural indication that the body is beginning to be worn out, and needs fresh material to repair its losses. And although the appropriation of the food is finally made by the

7

cells that compose the body, and so must be after it has been already digested and carried to them, nevertheless the sympathy of the different parts of the body with each other is such that hunger is satisfied by the mere act of supplying food to the stomach. Not only that, but the digestibility of the food has a great deal to do with it. Certain kinds of food, which we call rich, generally containing a great deal of fat and sugar, satisfy the hunger and produce a sense of satiety, when we have not really eaten enough to supply the bodily needs. This is because such food is digested very slowly, being so permeated with fat that the gastric juice, which does not digest fat, penetrates to the albuminous portions of the food with great difficulty. In such cases also fermentation frequently occurs, and persons who eat much so-called rich food may satisfy their hunger with it day after day, and still suffer from indigestion, and not get enough nourishment to repair the waste of the body. For these reasons, *plain food* is the best, especially for the young.

110. How much to eat.—The true way, therefore, is not to rise hungry from the table, but to stop eating when the hunger has been satisfied, and before any feeling of repletion comes on. It should be borne in mind that the process of digestion ought to go on without our consciousness. After a proper meal, the only sensation caused by the food we have taken should be that of complete satisfaction and contentment. If the stomach feels stuffed and full, we have eaten too much. It may be properly disposed of if the eater is in vigorous health, and able to rest for a time until the uneasy feeling of repletion wears away. But the whole

process ought to go on without causing us a moment's thought. If we are healthy, and if we treat
our digestive organs properly, we ought never to
feel that we have a stomach, or liver, or bowels.
They will never trouble us, if we do not trouble
them.

Our meals, therefore, should be sufficiently far
apart to allow an hour or two at least to intervene
between the digestion of one meal and the beginning of another. As digestion in the human being
ordinarily occupies from three to four hours, our
meals should be at least five hours apart, and this
is about the time usually allowed.

111. What to eat.—The matter of what to eat,
amid the great variety of foods, may safely be left,
in a healthy person, to the appetite. It is a familiar
proverb that " one man's meat is another man's
poison." Each individual must learn for himself
what food is the best for him. If any article is
found to disagree, it should thereafter be let alone;
no attempt should be made to overcome a natural
repugnance, and acquire an appetite for what is
distasteful. This is to fly in the face of Nature; it
is much the same as saying that one is competent
to direct the secret processes of nutrition and to
regulate the functions of organs, about which he
knows almost nothing, and which he can not control. Such action is intermeddling, not judicious
care.

112. Condiments. — Something should here be
said, however, about the use of certain substances
which are not foods, and yet are in common use
throughout the world, to make food more accepta-
ble to the palate. Such substances are pepper and

mustard. These condiments have two qualities that have caused them to be used in the preparation of food, viz., a peculiar flavor, which makes articles of food to which they have been added more savory, and a quality called pungency—i. e., they irritate any part of the body with which they are brought in contact. When either is placed upon the tongue, smarting is produced, sometimes to a painful degree, and tears start in the eyes. The effect can therefore be imagined when these substances are rubbed over the delicate mucous lining of the stomach during the movements of digestion. They can not but be extremely irritating, and therefore injurious. As a matter of fact, the excessive use of such things, whether alone or in highly-seasoned sauces (Worcestershire, etc.), results in extreme debility of the digestive apparatus and confirmed dyspepsia. The golden rule in the treatment of the stomach is, to put nothing into it that can be felt after entrance. As before stated, the operations of the stomach ought to go on without our consciousness of them. If enough spice is taken to produce a feeling of warmth in that organ, it is too much, and the mucous membrane has been irritated. We are all, or nearly all, born into the world with sound digestive organs, which need no spurring to make them do their duty. If they get out of order, it is our own fault, and rest will do more than anything else to set them right. If you whip a good horse, when he is doing his best, you will spoil him.

113. The Natural Drink.—The natural drink of all animals is water, for milk is to be looked upon as a food. Many people, however, are not satisfied

with water alone, but prefer it flavored with something else, and a great variety of drinks have been
invented. We shall consider here the effect that
some of these drinks, such as tea, coffee, and those
which contain alcohol, as malt and spirituous liquors
and wines, have upon the stomach.

114. **Tea and Coffee.**—Tea and coffee are found
to stimulate the nervous system, producing slight
exhilaration and relieving exhaustion without the
subsequent depression that follows the use of alcohol. In excess, however, they produce nervous
disorders, and probably all persons would be better
without them. Their influence upon the nerves,
the most impressionable part of a growing person,
renders them unsuitable articles of diet for the
young.

115. **Irritant Effect of Alcohol.**—If clear alcohol
is introduced into the stomach in small quantity, it
acts as an irritant, causing an increased secretion of
mucus. The mucous membrane lining the interior
of the organ becomes red and congested. If a large
quantity be introduced at once. the irritation of the
nerves of the stomach is so great that important
nervous centers are affected, and extreme depression immediately follows, soon ending in death.

116. **Effect of Diluted Alcohol on the Stomach.**
—The effects of diluted alcohol, as in the various
alcoholic beverages, differ in degree according to
the amount of alcohol they contain. A small amount
of alcohol, introduced into the stomach in this way,
produces slight congestion and reddening of the
mucous membrane, and excites the secretion of gastric juice, much as a small quantity of pepper might
do. As alcohol coagulates albumen, and when add-

ed to gastric juice outside of the body precipitates its pepsin and takes away its property of digesting food, it must be inferred that it will have a similar effect upon these substances when brought into contact with them in the stomach. This has been found to be the case when alcohol is taken in connection with food. As long as it remains in the stomach in any degree of concentration, the process of digestion is arrested, and is not continued until enough gastric juice is thrown out to overcome the effects of the alcohol. In large amount, alcoholic drinks often stop digestion entirely until the alcohol has been absorbed.

117. Effect of Large Amounts of Alcohol on the Stomach.—If alcohol be taken to the point of intoxication, the reaction after it has passed out of the body is very marked in the digestive organs. The pit of the stomach is tender to pressure, there are nausea and retching, with perhaps vomiting of mucus and bile. The tongue is heavily coated with fur, there is intense thirst, and entire loss of appetite. These symptoms indicate an acute catarrh of the stomach, and may not disappear for several hours.

118. Chronic Disease of the Stomach caused by Alcohol.—If the stomach is constantly being irritated by the habitual use of alcohol, serious changes result. The mucous membrane becomes permanently congested and covered with a thick, unhealthy, tenacious mucus, which sets up fermentation in the food, giving rise to what is called a sour stomach, heart-burn, and nausea, and sometimes vomiting of bile on rising in the morning. If warning is not taken in time, and the use of such drinks discontinued, the connective tissue of the mucous membrane becomes greatly developed, the walls of the stomach are thickened, the glands that secrete the gastric juice are pressed upon and many of them starved out of existence, so that the normal secretions that are necessary to digestion are scanty, and chronic and incurable dyspepsia results.

119. **Effect of Opium on the Stomach.**—Opium, when taken into the stomach, deadens the gastric nerves, and diminishes or entirely suspends the secretion of the gastric juice. It therefore interferes greatly with digestion. By deadening the nerves it diminishes hunger. It also in the same way prevents proper intestinal digestion (see next chapter), produces obstinate constipation, and thus seriously impairs the nutrition of the body. In many people the introduction of opium into the stomach is followed, after the narcotic effects have passed away, by nausea and vomiting and great prostration The same symptoms may follow the similar use of morphia. If the latter drug is administered by injection under the skin, so that it is immediately absorbed into the blood, the stomach symptoms may be entirely absent and the digestive powers undisturbed.*

In old opium-eaters this constant interference with the digestion and appropriation of food results in extreme emaciation, so that they may look almost like skeletons covered with parchment tight-

---

* The injection of morphia under the skin is sometimes followed by considerable irritation and even by a small abscess. If time is not allowed for the part to recover its natural condition before a second injection, inflammation is almost sure to follow. Hence the trunk and limbs of those who take morphia habitually in this way are generally covered with abscesses and the scars of those that have healed.

ly drawn over the bones. In some cases, however, even confirmed opium-eaters will recover a certain degree of appetite and digestive power, and the constipation be replaced by a diarrhœal condition.

120. Effect of Tobacco due to the Nicotine.— The effects of tobacco upon living organisms are mainly due to the nicotine it contains. This substance is a virulent poison, producing death, when given in a fatal dose, in less time than any other poison excepting prussic acid. During the operation of chewing or smoking it is absorbed by the mucous membranes, but is believed to be rapidly passed out of the body again, for there is enough nicotine in an ordinary cigar to kill the smoker if it were all taken into the system at once.

121. General Effect of Tobacco upon the Human Body.—When tobacco is taken by one who is unaccustomed to its use, or by any one in a poisonous dose, it irritates the mucous membrane of the mouth, causing an increased flow of saliva, soon giving rise to uneasy sensations in the stomach and bowels, nausea, vomiting, great distress at the pit of the stomach, a feeling of intense anxiety, giddiness, dimness of vision, and great feebleness and general prostration. The surface of the body becomes cold and clammy, the forehead is bedewed with perspiration, and, if the dose of tobacco has been sufficient, convulsions follow, and then death.

122. Effect of the Habitual Use of Tobacco upon the Digestive Organs.—Even in one habituated to its use, tobacco excites the flow of saliva, and has therefore been supposed by some to im-

prove the digestion. But the natural stimulant of
the salivary glands is food, and the outflow of saliva
in a healthy person is properly proportioned to the
quantity of food masticated, so that there is no need
of any artificial stimulus, and the excess of saliva
is harmful instead of beneficial. This is abundantly
shown by the dyspepsia that almost always affects
habitual smokers. Few of them can digest a meal
without a cigar, and few of them have a clean pink
tongue. It is almost invariably coated, a sure sign
that the stomach is in an unhealthy condition. It
is also probable that tobacco affects the nervous
supply of the stomach sufficiently to interfere to
some extent with the secretion of the gastric juice,
and with the muscular movements which are necessary to good digestion, though this can not be said
to have been proved. If it were only generally
known that the organs of a healthy body, such as
most of us start out in life with, need no prodding
to do their work satisfactorily, there would be less
sickness and longer lives.

The excessive secretion of saliva induced by the
use of tobacco is followed by dryness of the mouth
and throat, a natural result of the overwork forced
upon the salivary glands. This dryness leads in
many persons to the drinking of alcoholic beverages, water only satisfying the thirst temporarily,
while the alcohol keeps up the excitement of the
glands, which was started by the tobacco. The
excitement produced by the alcohol, on the other
hand, can be subdued to some extent by the seda-
tive tobacco. So these drugs play into each
other's hands, as it were, and keep their slave in
a sort of merry-go-round, the price he pays being

a loss of health and often of reputation and mental vigor.

123. Opium not a Food.—Although opium can not be regarded as a food in any sense of the word, it is habitually eaten by many people, particularly by Oriental people, as the Chinese and Javanese. It never acts, however, as a food, but, like other narcotics, it has the property of blunting the sensitiveness of the nerves and nervous centers, and thus it renders the sensations of hunger and fatigue less imperious, and enables persons to go longer without food or rest than they otherwise could. The lack of food and rest, however, is attended with great waste of unreplenished tissue, and consequent weakening of the vital forces.

124. Confectionery.—Confectionery is not injurious, when pure, unless taken in excess. Unfortunately, it is frequently adulterated, and, instead of containing simply sugar, flour, gum-arabic, and such harmless substances, is mixed with *terra alba* (gypsum), because it is heavy and cheap. Poisonous coloring-matters are also used. All candy that has a gritty feeling in the mouth should be rejected, and bright-yellow, orange, and green candies are to be looked on with suspicion, for they are almost always colored with chromate of lead.

125. Danger of Parasites in Food.—A word of caution is necessary about the eating of pork. This meat occasionally contains millions of minute parasitic worms, called the *trichi'na spira'lis*, and, if such meat is eaten without killing these worms, they are set free in the alimentary canal, bore their way into the blood-vessels, and are carried by the current of blood all over the body. When they come to vessels so small that they can not pass, they are stuck, dam up the blood-current, interfere with the circulation, and produce serious and often fatal disease. These parasites are killed by a temperature of 160° Fahr., and pork, therefore (including ham, of course), should never be eaten unless it is thoroughly cooked.* It is now known that many diseases are caused mainly by the presence of living microscopic organisms in drinking-water, typhoid fever, for example; and the immunity of the Chinese from this disease has been attributed to the fact that they only drink water that has been boiled, usually in tea.

Briefly, then, to keep the stomach healthy, *masticate the food thoroughly, eat when you are hungry, avoid overeating and eating between meals, eat plain food, do not spur the stomach with condiments or appetizers, shun alcoholic drinks, and use tea and coffee, if at all, with the greatest moderation and caution.*

---

* The flesh of the pig occasionally contains another parasite, called the *cysticer'cus cellulo'sæ*, which, if taken alive into the stomach, develops into the *tape-worm*. This parasite, like the *trichina*, is killed by thorough cooking.

CHAPTER V.

126. The Chyme.—After the partially digested food has passed out of the stomach into the intestine, it undergoes still further changes, and the difficulties of investigation in this part of the body are so enormous that very little progress has been made toward a clear explanation of what takes place there. Enough has been learned, however, to give us a general idea of how the process of digestion is completed.

We have seen that the *fats* and the *starches* are not digested in the stomach. The gastric juice does not act upon them at all, and they pass into the intestine in very much the same condition in which they enter the stomach. The fibers and tissues which hold the fats and starches together, being nitrogenous in their nature, are acted upon in the stomach and dissolved, so that the fat is set free and floats in globules like those upon the surface of a kettle of soup. The food thus prepared to pass into the intestine forms a thick, turbid, grayish fluid, called the *chyme*.

127. The Intestines.—The *small intestine*, into which the food passes from the stomach, is a tube about twenty feet in length, and an inch in diameter. It is composed, like the stomach, of three layers, the innermost one being mucous membrane, the middle one muscular fibers, some of which are circular and some longitudinal, and the outer layer serous membrane.*

The small intestine is connected with the large one by a valve-like opening situated in the vicinity of the right groin. The *large intestine* passes from this point upward to the liver, thence across to the left side, and then downward, constituting the last five feet of the alimentary canal.†

128. Muscular Fibers of Intestine.—The *muscular fibers* of the intestine contract with a worm-like motion, which always begins near the stomach, and extends slowly along the whole length of the intestine, gradually emptying it of its contents. In this

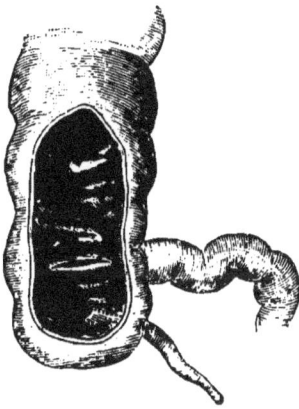

FIG. 27.—Junction of the small and large intestines, and the appendix vermiformis. The large intestine (here called the *cæcum*) is cut away so as to show the internal openings.

* The outer membrane of the intestine of animals, when separated from the rest, is used for sausage-casings, and, when properly prepared, also makes what is called gold-beater's skin.

† The beginning of the large intestine is situated in the right groin, and forms a sort of bag or pouch, called the *cæcum*. From one side of this pouch there projects a slender tube resembling the intestine in structure, and about six inches long. This is called the *appendix vermiformis*, i. e., the worm-like appendage (Fig. 27). In man it seems to be entirely useless, and is in fact a constant source of danger; for occasionally small objects, like cherry-pits and grape-seeds, which are swallowed with the food and not digested, become lodged in it, and gradually produce an irritation which results in an abscess, and destroys life. Such cases are not uncommon in medical practice.

slow passage of the food from the stomach through
the small intestine to the large one, it is mingled
with various fluids which complete the process of
digestion, and the nutritious portions of the mass
are absorbed and carried away by the blood and
other vessels.

129. The Duodenum. — The first eight or ten
inches of the small intestine are somewhat larger
than the remainder, and are called the *duode'num,*
because its length is about twelve fingers' breadth.
Into this duodenum empty small canals from two
very important organs, viz., the *pan'creas* and the
liver.

130. The Pancreas. — The *pancreas* (Fig. 28),

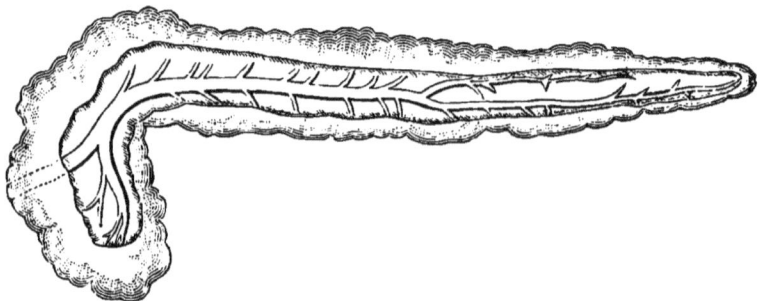

FIG. 28.—The pancreas, partly cut away, so as to show the duct, which
collects the pancreatic juice, and empties it into the duodenum.

which we call the *sweet-bread* when we cook it for
food,* is about six inches long, is shaped somewhat

* There are three kinds of sweat-breads, viz.: the thyroid-gland,
or throat sweet-bread, which is tough, almost like India-rubber; the
pancreas, or belly sweet-bread, which is more tender, and is quite commonly used; and the thymus-gland, or breast sweet-bread, which
exists only in young animals, wasting away as they grow up. This
gland is situated just behind the upper portion of the breastbone, attains its greatest size in human beings at the age of two years, and
disappears before the sixteenth year. Its use is not known. This

like a pistol, and is situated behind the stomach, with the large end, or the breech of the pistol, toward the right. It secretes a fluid, called the *pancreat'ic juice*, which has been shown to be the chief agent in the digestion of the *fatty* portions of the food. If a quantity of oil be shaken up with pancreatic juice, a white, opaque, creamy fluid is formed, in which the drops of oil or fat are not visible any more than they are in ordinary milk or cream. Microscopic examination, however, shows that the oil is not in any way decomposed, but is divided into very minute particles, in which condition it can be absorbed by the proper channels. In this way fat is taken up into the circulating fluids in its own proper form, and does not undergo decomposition until it reaches other parts of the body, if at all. The pancreatic juice also liquefies the nitrogenous matters which may have passed the pylorus undigested, and changes the starch into sugar. In fact, it seems to be the chief agent in completing the act of digestion, which has begun in the stomach.

131. The Liver. — The *liver* (Fig. 29) is a very large organ, the largest and heaviest in the body, weighing in a healthy adult from three to four pounds, and situated on the right side, protected by the lower four or five ribs. It secretes the *bile*, and from its size, and the amount of its secretion, is evidently one of the most important organs in the body, and yet its precise use is still a matter of dispute and doubt.

132. Liver-Sugar.—It was long supposed that the only function of the liver was to secrete the bile; but

gland, taken from calves and lambs, is the most tender and palatable sweet-bread of all.

it has been found, in recent years, that it also forms a kind of *sugar* in large amount. The blood which enters the liver is found to contain a small amount

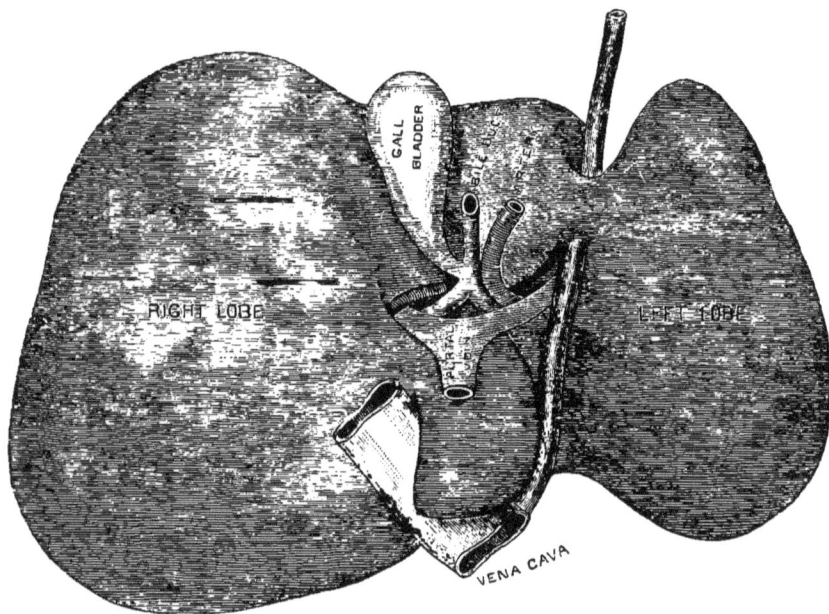

FIG. 29.—Under surface of the liver.

of sugar, while that which flows away from it, after having circulated through it, always contains sugar in considerable quantity. Even this fact, well established as it seems to be, is still a subject of dispute among experimental physiologists.

133. The Bile.—The *bile* is a somewhat glutinous fluid, of a rich, golden-red color,* which is discharged into the duodenum through the same open-

---

* When vomiting takes place and lasts for a time, the intestines reverse their action, and bile is carried backward through the pylorus into the stomach. It is here out of place, and produces extreme nausea. Its color is changed by the gastric juice to a greenish yellow.

ing with the pancreatic juice.* It must, therefore, become mingled with the food long before digestion is completed. The natural inference from this is that it has something to do with the process; but the digestion of every portion of the food can be accounted for in other ways. Nitrogenous matters are digested in the stomach, while the fatty matters and the starches are digested by the pancreatic juice, assisted, perhaps, by the intestinal juices, to be hereafter spoken of. It would appear, then, that there is nothing left for the bile to do, and that it must be an *excrementitious fluid*—i. e., that it consists of matters which have been separated from the blood by the liver because they are hurtful to the organism, and must, therefore, be expelled from the body. This was the ancient view, and it seemed to be supported by the fact that, if the liver be diseased so that this separation can not take place, and the constituents of the bile remain in the blood, *jaundice* occurs, and, if there is no relief, the person dies with all the symptoms of poisoning. So far, it seems plain enough that the bile has no office to perform in the body, but is only secreted to be expelled. But operations have been performed of such a kind that the action of the liver was not interfered with, and yet the bile could not enter the intestine, but had to be discharged outside the body through an artificial opening. Under such circumstances, if the

---

* As soon as the partially digested food, containing a certain amount of gastric juice, passes the opening of the bile-duct, there is a great gush of bile into the intestine. It is found that any acid, applied to this opening, will produce the same effect. The bile, being alkaline, neutralizes the gastric juice, which is therefore of no further use, and so the digestive process has to be completed by other means.

8

bile be simply an excrementitious fluid, its discharge from the body by one channel rather than by another ought not to make any difference in the health. But it is found, on the contrary, that operations of this nature are followed by every appearance of starvation. The appetite remains good, the digestion is not interfered with ; but, nevertheless, although food is supplied in abundance, extreme emaciation follows, and death generally in about a month. These facts show conclusively that the bile has some important part to play in the nutrition of the body.

It is found, moreover, by actual chemical examination of the excretions, that the bile, although it is discharged into the intestine, does not all leave the body. It must, then, be reabsorbed into the circulation. But, if this be so, why does it not give rise to symptoms of poisoning, just as if it were prevented from leaving the blood in the first place? The only possible answer to this is, that it is somehow changed in the intestine, so that when it is reabsorbed it is harmless.

134. The Intestinal Juices.—Besides the bile and pancreatic juice, the food meets in the small intestine with the *intestinal juices* proper. Of these very little is known with certainty, owing to the great difficulty of obtaining them from the living organism unmixed with other fluids. The small intestine is lined, however, with a mucous membrane containing millions of small tubules and glands, which secrete certain colorless alkaline fluids. Of these fluids it is both affirmed and denied that they possess the property of turning starch into sugar with great rapidity ; but, so far as is known, their part in the process of digestion is not important.

135. Absorption of Food.—If animals are killed at different times after the eating of food, and different portions of the intestine are examined, it is found that, while the upper portion of the small intestine contains a large amount of partially-digested food, the lower portion contains the shriveled remnants of ˉmuscular tissue, the husks of grains, the woody, indigestible fibers of vegetables, etc.; in short, the unappropriated residue of the food which has been taken. The great mass of what has been eaten has disappeared, and after a certain time the whole intestine will be found empty. There are two systems of vessels by which this absorption of food is accomplished—they are the *blood-vessels* and the *lacteals*.*

136. The Peritonæum.—To understand the arrangement of these vessels, it is necessary to know something of the *peritonæ'um.* The serous membrane, which has been spoken of as covering the outside of the stomach and intestines, covers to a greater or less extent all of the organs contained in the abdomen, and also lines the abdominal walls. This smooth, satiny membrane is called the peritonæum, and it renders the movements of the abdominal organs possible without discomfort to the rest of the organism. Now, the intestine being, as has been shown, a long, narrow circular tube, or canal, and the peritonæum passing entirely around it, there is a line running the whole length of the

---

* *Lacteals*, from a Latin word meaning milk, because when they are filled with the products of digestion they look as if they were filled with milk.

intestine, where the membrane becomes double, and this double fold is brought together like the gathers of a dress, and attached to the spinal column. So the intestine is loose in the abdomen, and still has an attachment to the spinal column. Between these two folds, or, in other words, within the double fold, between the two layers of membrane, the blood-vessels and lacteal vessels pass to the intestine (Fig. 30).

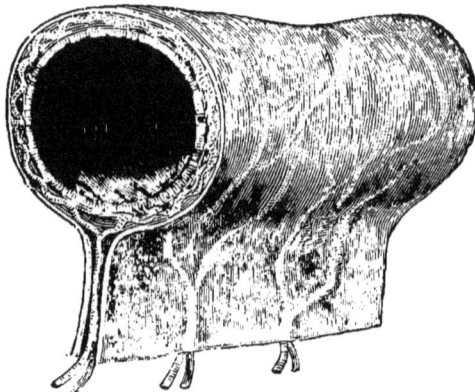

FIG. 30.—Diagram representing a cross section of the small intestine, showing the three layers, and the way in which the blood-vessels pass between the two folds of serous membrane (the peritonæum).

These vessels grow smaller and smaller and more and more numerous as they approach the intestine, and, when they at length enter its walls and penetrate to the mucous membrane, they divide into vessels so exceedingly minute as to be invisible to the naked eye, and fill the interior of the little projections of the mucous membrane, which are called *villi*.

137. The Intestinal Villi.—The *villi* are small projections on the surface of the mucous membrane, about a thirtieth of an inch long, and thickly covering the whole interior of the intestine, there being about ten thousand of them to the square inch, and about four million altogether (Fig. 31).

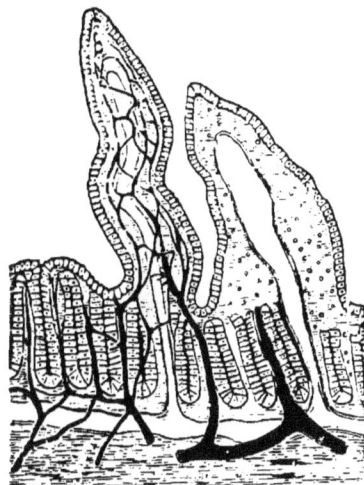

Each villus is covered with epithelium, and in its interior is a complicated mass of blood-vessels, twisted and knotted like a bunch of earth-worms (Fig. 32). In the very center of the whole is an open space, which is the commencement of a *lacteal.*

FIG. 31.—Section of the mucous membrane of the small intestine, showing two villi, and several secreting tubes or follicles; also lacteals, blood-vessels, and, at the bottom, the muscular layer.

138. The Lacteal Vessels. — The *lacteals* are only a part of a system of vessels, called the *lymphat'ics,* which extend everywhere throughout the body. The lymphatics all begin in a way that is not clearly understood, and gradually unite to form larger and larger vessels, until their contents are finally discharged into the veins and mingled with the blood. The fluid found in the lymphatics, called *lymph,* is yellowish, transparent, and saltish, and, being derived from the blood, nourishes the tissues, and takes up and carries away the waste. At certain intervals in their course, the lymphatic vessels are interrupted by small bodies called *glands,* * varying in size from a hemp-seed to

* The lymphatic glands are the bodies that sometimes undergo

an almond, into which the vessels enter, and from which they emerge. Whether they actually pass through the gland, or whether one vessel ends in it and another begins, is still a subject of discussion. But the lymphatic vessels all over the body have great absorbing power, taking up indiscriminately foods, poisons, or the waste of used-up tissues.

The *lacteals*, then, are that portion of the lymphatic system which is connected with the small intestine, and all the lacteals from the villi gradually unite to form a vessel called the *thorac'ic duct*, about as large as a goose-quill, which passes up close to the spine, and empties into a large vein very near the heart.*

FIG. 32.—Intestinal villus, showing the epithelial cells outside, the blood-vessels, and the beginning of a lacteal vessel.

139. The Portal Vein.—The blood-vessels which absorb the food from the intestines are veins, and they unite with the veins from the stomach, pancreas, and spleen, to form one large vein, called the *portal vein*, which enters the liver, so that all the blood from the digestive apparatus

slow inflammation in persons of a scrofulous tendency, forming hard lumps or abscesses in the neck.

* Many years ago, a man named Calvin Edson became extremely emaciated without any known cause. He was exhibited for a long time as "the living skeleton." After his death it was found that his thoracic duct was completely obstructed, so that none of the contents of the lacteals could pass into the blood. He died, therefore, of fat-starvation—i. e., a complete or almost complete deprivation of fat.

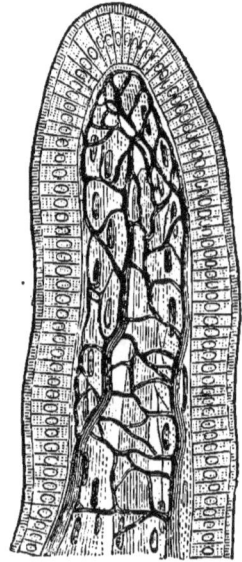

passes through the liver before it enters the general circulation.

140. The Chyle.—The villi, then, projecting as they do into the interior of the small intestine through its entire length, are continually bathed, during digestion, in the nutritious fluid which other organs have prepared for absorption. They float and sway about in this fluid, and suck it up as the roots of a tree get their sustenance from the soil,* and the blood-vessels probably have quite as much to do in the process as the lacteals. The latter absorb mostly the fatty matters in the cream-like form to which they have been reduced by the pancreatic juice. As the walls of the vessels are thin and transparent, the creamy contents, called the *chyle*, show through, and hence arises the white appearance during digestion which has given them the name of *lacteals*.

141. Changes in the Blood during Digestion.—

* The latest researches seem to show that the lacteal begins in the interior of the villus as a sort of hollow space, without any special wall of its own, and that around this space there are small fibers of involuntary muscular tissue. It is believed that, during the process of absorption, these muscular fibers contract at regular intervals. The effect of the contraction would be to pull down the top of the villus toward the base, and thus diminish the size of the hollow space above referred to, and empty its contents into the lacteal vessel. When the fibers relax, and the hollow space expands to its original dimensions, the fluid which has been forced into the lacteal is prevented from being sucked back again by the valves, with which all lacteal vessels are provided. The space is therefore filled again by the fluids which surround the end of the villus in the intestinal canal. In this way, by the alternate and regular contraction and relaxation of these minute muscular fibers, the villus acts like a suction-pump, and the intestine may be looked upon as lined with millions of microscopic suction-pumps, which work away during digestion, pumping the contents of the intestine into the lacteals, by which they are discharged into the blood.

As absorption goes on, the blood becomes more and more loaded with fatty matters, which can easily be recognized in it in the form of minute oily drops, but all of this blood passes through the lungs before it goes to the rest of the body. In its passage through the lungs, the fatty matters disappear in some way, not exactly understood, and the blood which comes away from the lungs contains none. After a time, however, as digestion progresses, the blood is so heavily charged with these oily matters that they can not all be decomposed, and a portion remains and is sent in the general circulation all over the body. If blood be drawn from a man or other animal at this time and allowed to stand, there will be a yellowish, creamy layer on its top.

Presently, however, the fat begins to disappear, as digestion approaches its close; the amount in the blood gradually diminishes, until it is entirely gone, the lymph in the lacteals becomes once more a transparent fluid, and digestion is complete.

142. Effect of Alcohol on the Liver.—Alcohol, being absorbed by the blood-vessels of the stomach, is borne away in the blood to the liver, and is brought directly in contact with its structure. If it is used habitually, though only in small quantities at a time, the liver may become the seat of serious changes. There may be a great increase of fat deposited in the cells, producing what is called by physicians the "fatty liver," or it may lead to a great increase of connective tissue between the cells and surrounding the blood-vessels. This newly developed connective tissue gradually contracts, and in so doing crushes the cells and obstructs the blood-vessels, making the organ much smaller than natural, and causing the surface to be covered with little projecting knobs, consisting of portions of liver-tissue that have been less compressed than the part that separates them. A liver thus affected is often called a "hob-nailed liver," from its appearance, or the "gin-drinker's liver." The pressure upon the liver-cells, and the destruction of many of them, prevent the proper formation of bile and liver-sugar. The substances composing the bile, not being separated by the liver with sufficient rapidity, accumulate in the blood and give rise to jaundice. The contraction of the newly developed tissue, by obstructing the blood-vessels, interferes with the circulation. The blood is dammed back, so to speak, the blood-vessels of the digestive organs are greatly distended, and the watery part of the blood oozes through the walls of the vessels into the cavity of the abdomen, causing dropsy. The distention of the blood-vessels of the stomach is often so great that some of the more delicate ones burst, and blood escapes into that organ. It is turned deep brown or black by the gastric juice, and produces nausea and vomiting of matters looking like coffee-grounds. This disease of the liver, called *cirrhosis*, is invariably fatal.

143. General Effects of Alcohol and Tobacco upon Digestion.—It will be seen from what has preceded that alcohol affects more or less almost every step of the digestive process. This interference with digestion, being in the direction of preventing the normal changes in the food before absorption, and also the absorptive process itself, powerfully affects the nutrition of the whole body. This effect is produced slowly, but is so marked as to be evident in the features of the drinker. The malt liquors seem to produce fatty degeneration, as a rule, while the stronger liquors, whisky, etc., cause the development of connective tissue. There is a lack of nutrition in both cases, but in the beer-drinkers the loss is masked by the growth of fat, while the spirit-drinkers become emaciated. Tobacco also often induces emaciation. The manner in which it is used does not seem to be important. Perhaps chewing interferes more with digestion, on account of the great increase of saliva, and the liability to swallow the tobacco-juice. The special dangers attributed to cigarette-smoking are probably not due so much to the quality of the tobacco or the wrapper as to the cheapness of the article, which leads to an inordinate consumption by very young persons.

144. The Spleen.—At the left extremity of the stomach, just under the ninth, tenth, and eleventh ribs, is an organ which is to this day a great puzzle to physiologists. It is called the *spleen*, and is about five inches long, four wide, and an inch thick. It is reddish in color, soft and pulpy in texture, with a very tough and strong fibrous covering. It receives its blood from a very large artery, and its vein, which carries away the blood from the organ, joins the portal vein, so that the blood from the spleen, like that from the other organs of digestion, passes through the liver, before it reaches the heart. The spleen is large in well-fed animals, and very small and shrunken in starved ones, while in some cases of disease, such as fever and ague, it reaches the enormous weight of twenty pounds, and forms an immense hard tumor in the left side.

The facts just stated would seem to imply that the spleen has some important office to perform in digestion, but what that office is no one has been able to discover.* It is a singular fact that the spleen may be entirely removed without permanent injury to the health. It has been several times removed from human beings, on account of disease. The effect upon them, although not so pronounced, is of the same character as the effect upon the lower animals, which show an enormous increase of appetite, usually gain considerable flesh, and acquire an unnatural ferocity of disposition. These things, however, do not seem to indicate that any particular function has been lost to the body, and the uses of the spleen are still a subject of earnest investigation.

* The most reasonable theory about the spleen at present seems to be that it has something to do with the destruction of old and worn-out blood-corpuscles and the formation of new ones.

CHAPTER VI.

THE BLOOD.

145. The Blood.—After the nutritious portions of the food have been taken into the blood, they pass through the lungs before they go into the general circulation. Before we consider the respiration, however, it is necessary to know something of the circulating fluid.

The *blood* is a thick, opaque fluid, varying in color in different parts of the body from a bright scarlet to a dark purple or even almost black. It has a somewhat viscid feel, a faint odor peculiar to itself, and a saltish taste.

146. The Red Blood-Corpuscles.—If a drop of blood be placed under the microscope,* immense numbers of small bodies will be seen, which are called the *blood-corpuscles* (Figs. 33 and 34). They are very minute, averaging only $\frac{1}{3500}$ of an inch in diameter, and are flattened in their shape. They may be described as looking like a cylindrical ring, the center of which has been filled up, but not quite to the level of the border, so that there is a slight depression on each flattened side. Taken singly,

---

* Prick the end of the finger with a pin. The most minute drop of blood is sufficient. Put it on a glass slide under a thin glass cover, and place it under a microscope.

these bodies, called the *red blood-corpuscles*, are of a light amber color, but in a large mass they give the characteristic red color to the blood. They vary somewhat in size in different animals, those of the monkey approaching most nearly to those of the human being. In birds, reptiles, and fish, they are very much larger, and instead of being circular are oval (Fig. 35).* They also have a distinct nucleus.

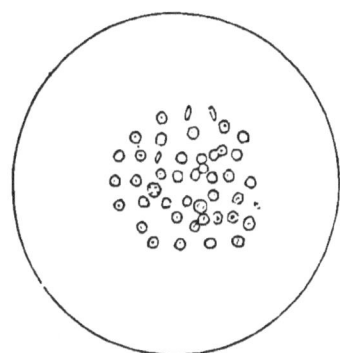

FIG. 33.—Human blood-corpuscles, including two white ones.

It is mainly by these microscopical differences that the blood of different animals can be distinguished

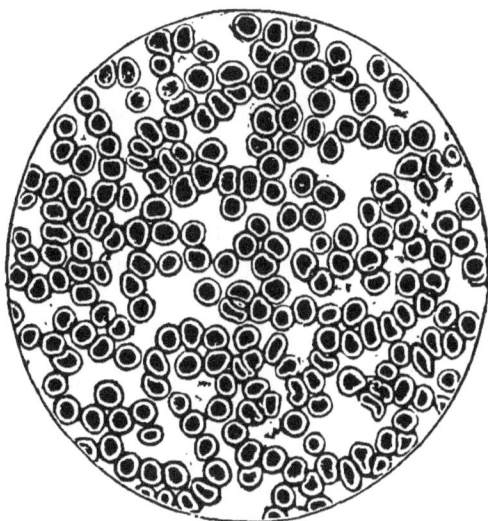

FIG. 34.—Human blood-corpuscles (highly magnified). From a photograph.

* The blood-corpuscles of the camel tribe are also oval, but smaller than those of birds.

from that of man, as is sometimes necessary in trials for murder.

147. The White Blood-Corpuscles.—The blood also contains *white* corpuscles in the proportion of one white corpuscle to three hundred red ones. They are larger than the red, are perfectly colorless, and globular in their form. The white corpuscles, under proper conditions, are seen to be continually changing their form, almost like living animals. There have been many speculations as

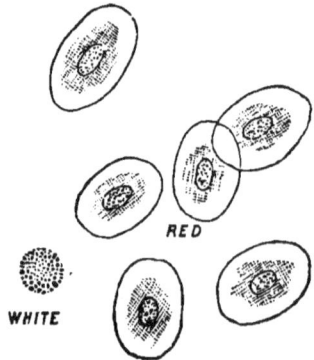

FIG. 35.—Blood-corpuscles of the frog.

to their office in the body, but nothing definite has been ascertained. In certain diseases, however, they are found to increase enormously in number, and some of these diseases are among the most dangerous and difficult to treat of any the physician meets with.

Although the blood-corpuscles are so very minute, they exist in such enormous numbers that they are estimated to compose half the mass of the blood.

148. The Plasma.—The fluid portion of the blood, in which these small bodies float, is called the *plasma*. It is almost colorless, quite transparent, and is nine tenths water. Its two most important ingredients are *albumen* and *fibrin.** The

* It is now believed that fibrin does not exist, as such, in fluid blood, but that there are certain substances (known as *fibrino-plastin* or *paraglobulin, fibrinogen, and fibrin-ferment*) which, by their inter-

former of these is chiefly concerned in nutrition, and the latter brings about the remarkable phenomenon known as *coagulation.* The plasma also contains various compounds of lime, soda, magnesia, etc., which have their own functions to perform in the nourishment of the body.

149. Coagulation of the Blood. — If blood be drawn from the living body, it very soon undergoes *coagulation.* This is due to its fibrin (or fibrin-factors); but the cause of the change is very imperfectly understood. After the blood has stood for a few minutes outside the blood-vessels, it at first becomes less fluid and assumes somewhat the appearance of jelly. Shortly the fibrin begins to contract and occupy a smaller space, gradually squeezing out the portion of the blood which still remains fluid. This mass which separates from the rest of the blood is called the *clot,* and the remainder, or still fluid portion, is called the *serum.* The contraction of the clot continues for several hours, until it forms quite a firm mass of a deep-red color, the remaining fluid being transparent and nearly colorless.

The red color of the clot is due to the fact that the red corpuscles become entangled in the coagulated fibrin, and, being semi-solid in consistency, they remain there and are not pressed out with the serum. It will be noticed that the *serum* is not the same as the *plasma.* The *latter* includes the *fibrin* (or fibrin-factors), while the *former* is *without it.*

150. Coagulation under Varying Conditions. — It is found that the blood coagulates more rapidly in

action, produce fibrin, and so cause coagulation. The evidence for this view is of too abstruse a nature to be given here.

thin layers than in a large mass; in a vessel or on a surface which is rough than in one which is smooth. For this reason the blood flows longer from a smooth cut in the body than from a wound with torn and ragged edges. In the latter case the blood coagulates very rapidly and stops the hemorrhage.

But the coagulation of the blood takes place not only outside the body, but, under similar circum-stances, inside, though not always with equal ra-pidity. If a vessel bursts inside the body, and blood escapes into the tissues around it, coagula-tion takes place after a short time; and this occurs even inside the blood-vessels, if there be any ob-struction to the circulation. If an artery or vein be compressed by a string or wire or finger, the blood will soon coagulate in the vicinity of the pressure. These facts have suggested the means used by surgeons to stop the flow from a bleeding wound, and will be referred to again.

151. Total Amount of Blood.—The *total amount of blood* in the human body is believed to be about *one twelfth* of the *whole weight* of the individual. Thus, in a man who weighs one hundred and fifty pounds there will be about thirteen pounds of blood, or somewhat more than a gallon and a half.

152. Oxygen in the Blood.—This rich, nutritious fluid is forced to all parts of the body in a way here-after to be described, carrying food to exhausted tissues and removing the used-up matters. A large part of the material necessary for the growth and nourishment of the body is taken in through the digestive organs; but there is a gas absorbed by the blood, in its passage through the lungs, which is even more necessary to life than food. This **gas**

is *oxygen*, which constitutes about one fifth of the atmosphere, and is essential to the life of all animals, probably without exception. We can live for days without food, but we can not live ten minutes without oxygen. Even water-animals are not exempt from this law; for fish extract the air, which is in solution in the water, by passing it through their gills. If a dish of water containing a fish be placed under the receiver of an air-pump, and the air be exhausted from it, the fish will be as surely drowned as a man would be if held under water. This is the reason why fish which are kept as pets in aquaria need fresh water continually. If a jet of water be kept falling into the vessel in which they live, so as to drag down bubbles of air, the water need never be changed, except for cleanliness.

153. Varying Color of the Blood.—The blood which enters the lungs is very dark, and sometimes almost black. When it has passed through the lungs, and flows away, it is of a bright scarlet. It has lost certain substances and gained others during this passage, and this beautiful and surprising alteration is produced by what is called the process of *respiration*.

154. Effect of Alcohol and Tobacco on the Blood.—Alcohol increases the amount of fat in the blood, exactly how is not known. It also lessens the capacity of the blood to absorb oxygen, and in this way interferes with the nutrition of the body. Tobacco is said to hinder the development of the red corpuscles, and in this way the pallor that so often accompanies the excessive use of tobacco is accounted for by some; this, however, has not been decidedly proved.

9

155. Respiration a Complicated Process. — At first sight, the process of respiration is a very simple one, consisting merely in the inspiration and expiration of air. In reality, however, it is complicated—certain very essential parts of it going on without our consciousness, like so many of the phenomena of digestion. The organs mainly concerned in the acts of respiration are the lungs; but there are certain additional organs, whose functions, if not absolutely necessary, are certainly important.

156. The Nasal Passages. — The air, before reaching the lungs, goes through several passages, lined throughout with mucous membrane. The human being can breathe through either the nose or the mouth ; but in some animals (the horse, for instance) respiration only takes place through the nose, and, if this be closed, suffocation follows. The external openings of the nose are guarded by short, stiff hairs, which grow just inside the nostrils, and which serve to purify the air somewhat, as it passes through them, by catching and retaining particles of dust. The interior of the nose is so formed that it is not a large, open, free passage, but has a number of projecting bones, running lengthwise along its walls, which are covered with moist membrane and present an extensive mucous surface to attract particles from the air. If we breathe through the mouth, on the other hand, the air goes directly to the throat, and the cavity of the mouth is so large that the purifying effect of the moist membrane is hardly perceived. This shows how much better it is to breathe always through the nose; for the air, undoubtedly, in this way, is rid of many impurities; and physicians habitually, and almost unconsciously to themselves, keep their mouths shut as much as possible when they are exposed to a contagious disease.

From the nose the air arrives at the throat, and thence it passes into the windpipe, or trachea, through a small opening called the glottis.

157. The Trachea.—The *tra'chca* (Fig. 36) is a tube about four and a half inches long and an inch wide, which divides at its lower extremity into two smaller tubes called *bronchi*, one of which goes to each lung. It is mainly fibrous in its structure, and it is kept open to its full extent by a number of rings of cartilage, placed at a short distance apart through its whole length. The trachea is situated in the neck just in front of the œsophagus, and as these stiff rings might press backward on the œsophagus, and thus interfere with the process of swallowing, they do not pass completely around the trachea, but are lacking in the part next the œsophagus, comprising about one third of the whole circumference of the tube. At the upper extremity of the trachea is the *lar'ynx*, or the organ of voice, which is essentially a triangular-shaped box of cartilage, the lower end opening freely into the trachea, and the upper being closed by muscles and membranous tissues, with the exception of the opening of the glottis.

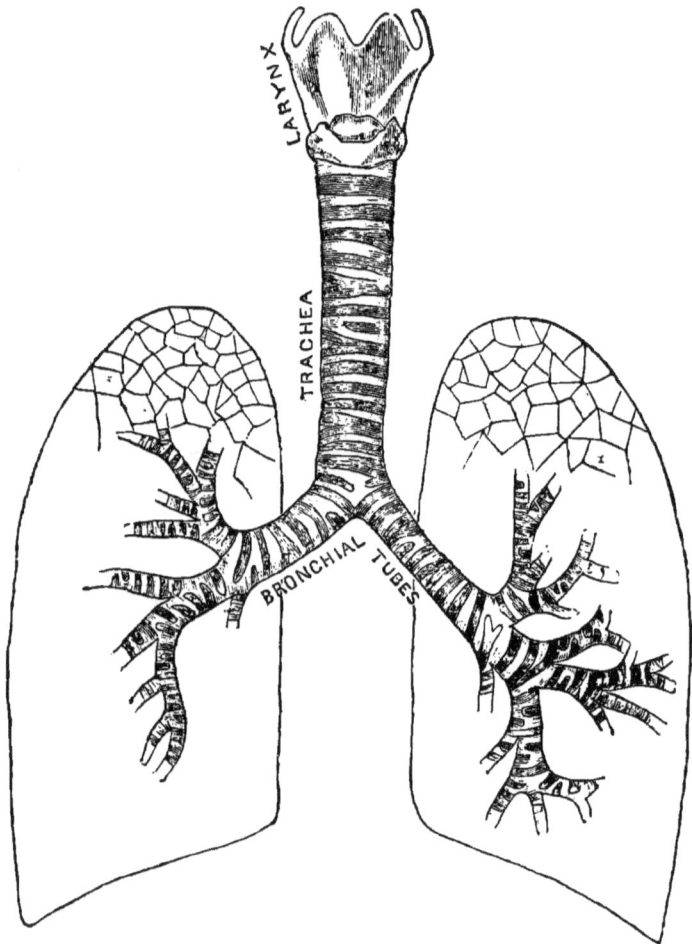

FIG. 36.—Larynx, trachea, and bronchi, showing the manner of division, and the rings of cartilage.

158. The Glottis.—The *glottis* is a slit-shaped opening, a little less than an inch long, extending from before backward and from above downward, not being, in other words, either perpendicular or horizontal in the throat, but shelving toward the rear. The front extremity is at the base and back of the tongue, and the opening itself is bounded at the sides by two firm, fibrous, strong, pearly-white membranes, called the *vocal chords*, by the vibration of which sound is produced. These vocal chords can be separated to the extent of half an inch, or brought together so as to touch, by the muscles which are attached to the back part of the larynx. The production and modulation of the voice will be treated of hereafter.

At the base of the tongue, springing upward just above the forward end of the glottis, is a stiff piece of cartilage, shaped like a leaf with a rounded end. This is called the *epiglottis*, and probably performs two functions, viz., that of protecting the glottis from food or other substances during the act of swallowing, and that of directing the column of expired air up toward the roof of the mouth or throat, and so aiding in the modulation of the voice.

159. The Lungs.—The essential organs of respiration are the *lungs*, which are two in number and fill nearly the whole cavity of the chest, a portion, however, being occupied by the heart and large blood-vessels. The lungs are very light in proportion to their size, and in animals they are commonly called "the lights." They weigh together only about two .pounds and a half, and easily float in water. In small children they are of a beautiful pinkish color, but in older persons they become slate-colored, and have black spots scattered here and there over their surface.

160. Minute Divisions of the Lungs.—After the trachea divides into two bronchial tubes, one of which goes to each lung, these bronchial tubes continue to subdivide into smaller and smaller tubes, all the branches diverging widely from each other, until their diameter is diminished to about $\frac{1}{20}$ of an inch. At about this point the cartilage rings disappear, but the tubes still divide until the smallest are only $\frac{1}{50}$ of an inch in diameter. At the very ends of the smallest tubes, there is an enlargement about $\frac{1}{12}$ of an inch in diameter, called a *pulmonary lobule* (Fig. 37). It constitutes a small cavity, into which dip little partitions, that do not meet each other, but create minute hollow spaces around

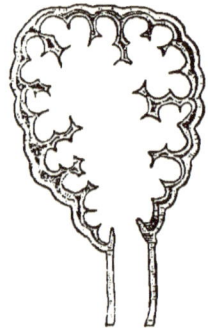

FIG. 37.—Section of a pulmonary lobule, showing its division into pulmonary vesicles.

the sides of the lobule, called *pulmonary vesicles.* These are about $\frac{1}{75}$ of an inch in diameter, and are the smallest divisions of the lung.

FIG. 38.—Ciliated epithelium from a small bronchial tube. The small round cells at the bottom are young ones.

161. The Lining Membrane of the Lungs.—All these tubes and passages, down to the most minute, are lined with a delicate mucous membrane, which has this remarkable peculiarity. The little epithelial cells with which all mucous membranes are covered, have in this situation what are called *cilia* at their ends (Fig. 38). That is to say, each cell has at its tip a fine, hair-like lash, which keeps in constant motion, as long as the person lives, and for some time after he is dead. If a piece of the mucous membrane from the throat of a frog just killed be snipped off with a pair of scissors and placed under the microscope, this incessant motion of the cilia may be easily seen. Although they are so delicate in their structure, they are so innumerable and act in such perfect concert, that they keep up a constant current toward the outside of the body. They probably aid in the expulsion of the foul gases which the blood leaves in the lungs.

162. Effect of Alcohol and Tobacco on the Nose and Throat.—The habitual use of alcoholic drinks often results in a chronic catarrh of the throat and nose. The membrane becomes red, swollen, streaked with enlarged veins and covered with stringy mucus, which the drinker is constantly trying to get rid of by hawking and spitting. This chronic inflamed condition of the throat produces almost constant thirst, generally gratified by more alcohol, which keeps up the irritation it is meant to soothe. This inflammation often involves the vocal chords, giving rise to the well-known husky voice of the old toper.

Much the same condition of the nose and throat is brought about by smoking tobacco, the smoke constantly irritating the mucous membrane of the throat and nose, and producing chronic catarrh of those parts. It is said that no habitual smoker has a healthy throat.

The excessive secretion of saliva in those who chew tobacco produces extreme thirst, and may thus lead to the habitual use of alcoholic liquors.

163. Effect of Alcohol upon the Lungs.—The effect of alcohol upon the lungs of habitual drinkers is very marked. The mucous membrane becomes congested and the secretion of mucus more plentiful. Colds are easily contracted and hard to get rid of, and a chronic bronchitis results. This finally diminishes the elasticity of the lung tissue, and the air-vesicles become in spots permanently distended like little bladders. The function of respiration is thus interfered with, and the sufferer becomes short of breath, often asthmatic, is troubled with a constant cough and hawking of mucus, so that his life is a burden.

Alcohol, instead of preventing consumption, as was once believed, reduces the vitality so much as to render the system unusually susceptible to that fatal disease.

164. Asthma. — The smaller bronchial tubes, which have no rings of cartilage, are nevertheless surrounded by involuntary muscular fibers. When, in consequence of disease, these fibers contract strongly, they diminish the caliber of the tubes, and render it very difficult sometimes for the sufferer to get air through them in either direction. This condition gives rise to great distress and a sense of suffocation, and is called *asthma.*

165. The Blood-Vessels of the Lungs.—Between the pulmonary vesicles run the small blood-vessels immediately under the delicate mucous membrane, so that the blood comes almost in contact with the air that we breathe. They surround the vesicles completely, and it is in this part of the lung that the great changes take place in the blood during respiration.

166. The Outer Covering of the Lungs.—The outside of the lung is covered by serous membrane, and so is the inside of the chest-wall. This renders the movements of the lung painless and easy. This membrane is called the *pleura*, and when it becomes inflamed, in the disease known as *pleurisy*, respiration becomes excessively painful.

167. Inspiration.—As the cavity of the chest is enlarged, the air already in the lungs is rarefied, and the external atmospheric pressure forces air in to fill the organs. We have already stated that the ribs are so shaped, and so connected with the spine behind and the sternum in front, that when they are raised up toward the shoulders the sides move outward, and the sternum moves forward. . This motion of the ribs is caused partly by powerful muscles attached to their external surface all the way down the chest, and partly by short muscles which pass between the ribs from the lower edge of each one to the upper edge of the one just below it.

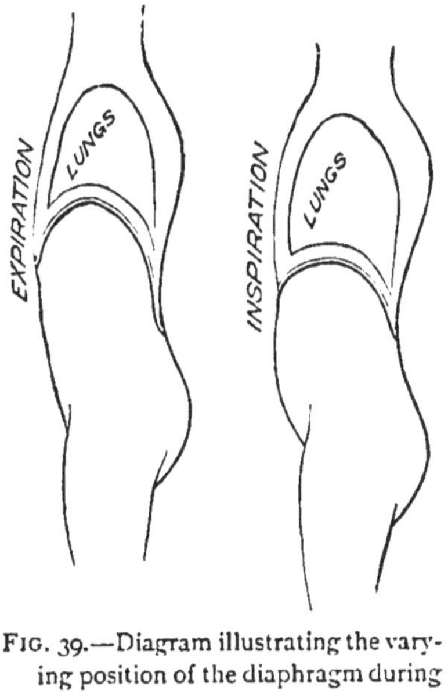

But, in addition to its expansion toward the front and sides, the cavity of the chest is enlarged in a downward direction by the contraction of the *diaphragm* (Fig. 39). This muscle

FIG. 39.—Diagram illustrating the varying position of the diaphragm during respiration.

has a strong, flat, tendinous center, from every side of which strong muscular fibers pass to the walls of the chest. It separates the chest from the abdomen, and while the muscular portion of it is attached to the lower ribs, the spine and the very end of the breastbone, the center rises much higher in the chest, so that it has the shape of a vaulted roof, on top of which are the lungs and heart, and underneath the stomach and liver. Of course there are passages through it for the blood-vessels and nerves, but these openings are so guarded that the diaphragm forms a tight partition. Now, as the center of the diaphram rises so much higher than the sides, it is very evident that a contraction of the muscular fibers will pull the center downward, and so increase the capacity of the chest. And this is what actually occurs at every inspiration. When this contraction takes place spasmodically, the air is drawn into the lungs with a sudden impulse, and we call it *hiccough.* *

168. Expiration. — Inspiration then involves a contraction of many muscles, and they act with a great deal of force, for they have to lift the atmosphere, which is pressing on the outside of the chest with a force of fifteen pounds to the square inch. By this simultaneous contraction, the ribs are drawn out of their natural position—i. e., they are drawn upward into a position which they never would assume if left to themselves. By the elasticity of their cartilages and other tissues attached to them, they tend to return to their former position as soon as the force which has drawn them out of

---

* *Hiccough*, being due to a spasmodic action of the diaphragm, may be stopped by any means that tends to control the spasm. The easiest method is to put the diaphragm on the stretch, as follows : prolong the act of expiration as much as possible, and at the end make a forcible expiration ; then inspire slowly and take as full an inspiration as possible. It is rare that a second trial will be necessary.

it ceases. The diaphragm, also, when contraction stops, tends to recover its former arched shape. The lungs also contain, in addition to the elements already mentioned, a large amount of elastic fibers, interlaced with the other tissues in every direction. These, too, as soon as the pressure which has stretched them ceases, tend to return to their former condition. This elasticity of the different organs concerned in the act of respiration, then, brings the chest and lungs back to the condition in which they were before inspiration began. This is the ordinary act of *expiration.*

169. Relative Force of Inspiration and Expiration.—As we usually breathe, then, the act of inspiration is an active one, requiring effort and powerful muscular contraction, while the act of expiration is passive, and is accomplished by the elasticity of the tissues.*

Under other conditions, however, the act of expiration may be more powerful than that of inspiration. There are strong muscles connected with the chest in such a way as to act in opposition to the muscles of inspiration, and make the cavity of the chest smaller than it ordinarily is. It is by the active contraction of these muscles that we produce what is called a forced expiration, which has been estimated by careful observers to be one third more powerful than a forced inspiration.

170. Amount of Air respired with Each Breath.

* The outer surface of the lungs is kept in contact with the chest-walls by atmospheric pressure. If the chest-wall be punctured, so that the air-pressure is the same both outside and inside of the lung, the elasticity of the organ is such that it immediately collapses, driving out all the air from its interior.

—The *amount* of *air* taken into the lungs with each inspiration is about *twenty cubic inches.* Now, the entire capacity of the lungs varies in different persons from one hundred and fifty to two hundred and fifty cubic inches or even more. So that with each breath, a very small amount, generally not one tenth, of the air in the lungs is changed. It is even estimated that after the most forcible expiration possible, at least one hundred cubic inches of air will remain in the chest of a man of medium size, which can not be expelled. In ordinary breathing, therefore, only the air in the larger bronchial tubes can pass in and out of the lungs. But the changes in the blood must be produced at the extreme end of the finest tubes in the pulmonary vesicles. So the question arises, How does the air get to the vesicles?

171. How the Air in the Lungs is changed.—In the *first* place, the law of the *diffusion of gases* comes in play. When two gases come together, they tend to mingle with each other until they finally occupy equally the whole of the vessel or other confined space in which they may be. After the mixture, each gas will be found in the same proportion in every part of the vessel. Now, the air in the pulmonary vesicles and smallest bronchial tubes is heavily loaded with carbon dioxide (carbonic acid), while that which is drawn in with inspiration is rich in oxygen. These two gases, then, carbon dioxide and oxygen, are constantly being diffused throughout the whole of the lungs. In the *second* place, the cilia, which have already been described, being in constant motion, keep up a current of the foul air from the pulmonary vesicles along toward the larger bronchi and trachea, and fresher air keeps constantly pressing in to fill the place of what has been in this way removed. Thus, in the smallest bronchial tubes, there are always two currents of air passing each other in opposite directions : one, immediately next the mucous membrane, being a thin layer moving outward ; and the other, in the center of the tube, moving inward (Fig. 40). So that the air in the larger bronchi and trachea is changed periodically by the acts of inspiration and expiration, while the circulation of the air in the small bronchial tubes and pulmonary vesicles is continuous.

FIG. 40.—Imaginary section of a small bronchial tube, showing the influence of the cilia in producing an outward current of air.

172. Amount of Air respired daily.—The amount of air taken in with every inspiration is about twenty cubic inches. The average number of respirations per minute is eighteen. This is a matter which varies very much with the individual. Children and women breathe somewhat more rapidly than men ; but taking eighteen as the average, the quantity of air breathed per minute is three hundred and sixty cubic inches, or about one fifth of a cubic foot. In an hour, then, we use about twelve cubic feet of air, and in a day nearly three hundred cubic feet. This amount is increased by every muscular exertion, and also by the curious fact that the ordinary respiration does not seem to be altogether sufficient for the needs of the body, and every now and then we draw a deeper breath than the average. This occurs usually about once in every five or six acts of respiration. Considering the increase in the amount of air respired at each long breath, and the increase of rapidity of respiration due to slight causes during the day, it is estimated that an adult really respires about three hundred and fifty cubic feet of air per day.

173. Changes produced in the Air by Respiration.—When the air enters the lungs it contains nearly 21 per cent of *oxygen* and 79 per cent of *nitrogen*, with about one twentieth of one per cent of *carbon dioxide*, a little *watery vapor*, and a trace of *ammonia*.

If the air be collected at expiration, after having undergone the changes in the lungs, we find the following:

1. *It has lost oxygen.*
2. *It has gained carbon dioxide.*
3. *It contains more watery vapor.*

The watery vapor in the expired air is not ordinarily visible, but in cold weather, when it becomes condensed, it can be very plainly seen. The whole amount of water passed away daily in the breath of a man has been carefully estimated, and found to average about one and one sixth pound avoirdupois.

174. Former Theory about the Formation of Carbon Dioxide.—Out of the four cubic inches of oxygen taken into the lungs with each inspiration, one cubic inch disappears. The *carbon dioxide* which is exhaled from the lungs consists of carbon and oxygen united in certain proportions, and it used to be supposed that the carbon in the blood united with the oxygen of the air in the lungs themselves, forming carbon dioxide, and that in this way the carbon, released by the wear and tear of the body, was got rid of. Now, the process of *combustion* in a flame of any kind consists in this same change, viz., the union of the carbon and hydrogen of the oil or other inflammable substance with the oxygen of the surrounding air, forming carbon dioxide and water, and giving out heat during the process. So it was for a long time thought that the lungs were a sort of furnace in the body, where the carbon and hydrogen of the blood were burned, so to speak, and the products of combustion exhaled, while the heat occasioned by the process kept up the warmth of the body. This was a beautiful theory, but it is found not to be warranted by the facts.

There is *more oxygen absorbed* in the lungs, with every respiration, *than is exhaled* in the carbon dioxide and watery vapor taken together. This fact of itself disproves the above theory, for it shows that a portion of the oxygen disappears in the lungs, or is carried away by the blood.

175. Organic Matter in the Breath. — Besides the carbon dioxide given off in the expired air, there is a certain amount of organic matter, containing nitrogen, which gives the breath a slight but peculiar odor. Where many persons are breathing in a badly ventilated room, this organic matter accumulates, and imparts to the atmosphere that odor which we all recognize as peculiarly oppressive and close.

176. Changes in the Blood during Respiration. —The blood undergoes changes in its passage through the lungs which correspond to the changes in the air. In the first place, it is altered in its color. As it enters the lungs, it is of a deep bluish purple, almost black; as it emerges, it is of a beautiful and most brilliant scarlet. On chemical examination, to determine the cause of this remarkable change, it is found that the blood which comes away from the lungs contains more oxygen and less carbon dioxide than that which enters them. Additional proofs that the formation of carbon dioxide does not take place by direct combination in the lungs are the facts that the venous blood, before it enters the lungs, is deeply charged with carbon dioxide already formed, and that the blood which comes away from the lungs contains oxygen in free solution.

The brilliant color, which is the result of this change in the blood, has not yet been satisfactorily accounted for. It has been proved that the oxygen and carbon dioxide are carried by the blood-corpuscles, and not by the plasma, and the change of color in the blood is entirely due to the change in those minute bodies. They have been said to change their shape and become more globular in one case than in the other, but the attempts to explain the difference of color have not yet been entirely successful.*

177. Where the Carbon Dioxide is formed.—If the carbon dioxide is not formed in the lungs, then where does it come from? Experiments of the most ingenious kind have been performed to determine this question, and they are too long to

---

* The coloring-matter of the red corpuscles is called *hæmoglobin.* It is found that this substance, when united with an excess of oxygen, forming oxyhæmoglobin, has a bright scarlet color, and, when the amount of oxygen is greatly reduced, is of a dark purple. But this does not explain much.

mention in detail. But it has been conclusively shown that most of the carbon dioxide is formed in the tissues in all parts of the body, during the processes of nutrition. And even here it is not produced by a direct combination of the oxygen with the carbon, for the exhalation of carbon dioxide will continue for a considerable time in an atmosphere of hydrogen, where of course there is no oxygen furnished to the tissues. The carbon dioxide, then, is formed by decomposition of the tissues, and the oxygen is used by them to build themselves up again.* The amount of carbon dioxide given off in the breath has been found to be somewhat less than one cubic inch, or about fourteen cubic feet per day, weighing about a pound and a half, and representing waste of the organism to about this amount.

178. Effect of Alcohol and Opium on the Function of Respiration.—Although the evidence is conflicting, it is now generally conceded that alcohol diminishes the excretion of carbon dioxide and watery vapor from the lungs. Opium has the same effect. It is probably the result of diminished activity of the cells, due directly to the paralyzing effect of these drugs.

Opium lessens the frequency of respiration and renders the throat and mouth dry by diminishing the secretion of mucus. In fact, it diminishes all the secretions of the body, excepting the perspiration, which is increased by it.

179. Composition of Air.—Air being so essential

---

* It will be understood that the place of the carbon which is lost to the body in the carbon dioxide which passes off by the lungs, is supplied by the fresh material taken in with the food.

10

to life, it is evidently important to have it as pure as possible. It must contain enough oxygen, so that with each respiration the temporary needs of the body may be satisfied, and should contain no substances which are injurious to life or health. Now, the air normally contains about four parts of nitrogen to one of oxygen, and the variation in the amount of these constituents is found to be surprisingly small in widely different localities.* It always contains a small amount of carbon dioxide, and a variable quantity of watery vapor.† It is also never found

* The following analyses, mostly by Mr. Angus Smith, show this very clearly :

Locality whence air was taken.	Percentage of oxygen.
Middle of Hyde Park, London	21.005
Sea-shore and heath, Scotland	20.999
Tops of hills, Scotland	20.98
Forests, Scotland	20.97
Summit of Mt. Blanc	20.963
London, open parts, summer	20.95
Bottom of hills, Scotland	20.94
Open parts of Glasgow	20.929
Marshy places	20.922
Chamonix, Switzerland	20.894
Sitting-room, feels close	20.89
Air procured from balloon 18,000 feet high	20.88
Theatre gallery, 10.30 P. M	20.86
In small room, kerosene-lamp burning	20.84
In small room, after six hours	20.83
Theatre pit, 11.30 P. M.	20.74
Closer parts of Glasgow	20.706
Mines, under shafts, average of many	20.42
Mines where candles go out	18.50
Very difficult to remain in	17.20

† This watery vapor is a very necessary constituent of the air. Out-of-doors the amount of it is regulated in ways beyond our control ; but in-doors, unless special care is taken, the air may be so dried by artificial heat, that when respired it will absorb more than the ordinary amount of moisture from the mucous lining of the lungs. Then the

entirely free from impurities, such as other gases
than those named, in small quantity, and minute
floating particles of matter, which we group together under the common name of dust.

It has been shown, however, that the breathing
of animals is continually removing oxygen from the
air and increasing the amount of carbon dioxide.
Now, carbon dioxide is a poison to animals, and
if inhaled in large amount produces almost immediate unconsciousness and death. It is for this reason that it is being constantly rejected from the
body. If this process of removing oxygen from
the air and adding carbon dioxide to it were to
go on indefinitely, it is evident that after a time
the one would be so much reduced in amount,
and the other so much increased, that animals
would die of carbon-dioxide poisoning—i. e., of
asphyxia.

180. Respiration of Plants. — This danger is
guarded against in the outer atmosphere by the
constant absorption of carbon dioxide by plants.
All plants, through their leaves, decompose carbon dioxide into its original parts, carbon and
oxygen. The carbon they appropriate for their own
nourishment, and the oxygen they return to the atmosphere. Thus the respiration of plants is exactly
the reverse of that of animals. The latter absorb
oxygen and give out carbon dioxide, and the former absorb carbon dioxide and give out oxygen.

mucous membrane becomes dry, there is an increased flow of blood to
the part, and, if the dryness of the air is not remedied, inflammation
may result—i. e., a catarrh. For this reason a vessel of water should
always be kept on the top of a heated stove or furnace, that its evaporation may insure sufficient moisture in the air to prevent injury to the
lungs and throat.

By this never-ending interchange the proportions
of oxygen and carbon dioxide in the atmosphere
are kept about the same.

181. Contamination of the Air in Houses.—In-
doors, however, there is no opportunity for this
self-purification. Even if a few plants are kept in
the house, the amount of carbon dioxide they con-
sume is very little, and the effect they are able to
produce toward purifying the room can not be compared with that of the immense stretches of forest
and plain out-of-doors. Moreover, the amount of
carbon dioxide in houses is increased by combustion. A five-foot gas-burner throws out as much
carbon dioxide as five men. The unhealthiness of
a closed room is also increased by the organic matter of the breath, which is very poisonous.* The
odor of this matter is perceptible in a room long
before the accumulation of carbon dioxide reaches
a point when it is likely to be injurious. It is, therefore, to be looked upon as by far the most dangerous impurity in the atmosphere of an occupied
room.

182. Ventilation.—In order to forestall any evil
result from such impurities, the air of a room should
be changed frequently enough to prevent the odor
of this organic matter from being perceptible. This
usually requires some special attention, and is called
ventilation. In warm weather, all that is necessary
is to open the doors and windows and allow the air
to circulate freely through the house. But in cold

---

* The composition of this organic matter is not known. It is given
off in such small quantity that the chemists have never been able to
analyze it. It putrefies rapidly after it has left the body, and then becomes very offensive.

weather more care is required. A fireplace, with an open fire, is an excellent means of drawing out the foul air—sending it up the chimney, and so out of the house. The fresh air, to supply the place of what has been thus removed, may come in through cracks in the windows and doors. But the fresh air admitted in this way in cold weather, being heavier than warm air, falls and sweeps along the floor. This is very dangerous, for few people can endure a cold draught on the feet and ankles, while the rest of the body is warm, without taking cold. Moreover, the smallness of the apertures through which the air comes increases the rapidity of the current. It is better, therefore, to let in the fresh air through a special opening, so arranged that the cold air shall not immediately fall to the floor. This can be done cheaply and effectively by raising the lower sash of the window about four inches, and putting underneath it a board, fitted to close the opening tightly between the sash and the sill. There will then be a long, narrow opening between the upper and lower sash, through which air will enter in a current directed upward toward the ceiling, and, before it descends, its momentum will be so much diminished that it will not create a draught. In very cold places, where double windows are used, the same result may be obtained by raising the lower outer sash a little, and lowering the upper inner one. The best way, however, is to warm the fresh air before it enters the room; but this is too large a subject for discussion here.

This foul organic matter from the lungs of animals, when it gets out into the open air, is immensely diluted, and, being acted upon by the oxygen of the atmosphere, is changed into other and less harmful substances, which, in their turn, are washed down by the rain and become a part of the soil.

183. Contagious Diseases. — The air is not only polluted by these products of the respiration of healthy animals, but it is made unfit for breathing, in a way involving still more danger to life, by the matters given off from the lungs and bodies of sick persons. There are certain diseases which are called *contagious* or infectious, because they can be communicated from one person to another. Such diseases are small-pox, measles, scarlet fever, typhus fever, diphtheria, and perhaps consumption.* It is known that the matters contained in the air expired from the lungs, or, in some cases, specks of matter cast off from the skins, of persons sick with these diseases, will produce similar diseases in persons who inhale them. Exactly what it is that reproduces the disease is not known, but there is believed to be a *microscopic organism,* peculiar to each disease, which, like a kind of seed, will always produce that disease by its own growth and multiplication whenever it meets with proper conditions. Whether these little organisms ever grow and multiply outside of the body we do not know, but that they do so in the blood we have abundant evidence.†

* Whooping-cough, mumps, and chicken-pox, are propagated in a similar manner, but are less dangerous.

† It is thought by some that malarial fevers (fever-and-ague, etc.) are produced by microscopic organisms of this kind, but this is uncertain. It is well for those who live in districts where such diseases are prevalent to remember that the poison, whatever it may be, is most active in the spring and fall, at night, and near the surface of the ground. In such a region, and at such a season, therefore, people should not go

There are other diseases which are believed to be produced by similar organisms growing and multiplying in the discharges from the stomach and bowels. Such organisms are believed to grow outside the body as well as inside, and are supposed to be the cause of Asiatic cholera, typhoid fever, and yellow fever.*

184. Precautions against such Diseases. — It is probable that all of these microscopic organisms (called bacteria, bacilli, micrococci, etc.) which float about in the atmosphere, if they do not find a favorable place in some animal body, where they can grow and propagate their kind, finally die. If it were not so, the human race would be exterminated by them. But men have two ways of dealing with them so as to prevent their spreading. One is to separate the sick person from well ones, as far as possible, and the other is to kill these little organisms as fast as they leave the body and before they can get out of the room. This is accomplished by the use of powerful drugs, called *disinfectants.*

Nurses and doctors adopt special means of warding off infection, or are willing to expose themselves

out after sundown, should keep their bedroom-windows closed, and should sleep above the first story.

* The discharges from the bowels and kidneys of healthy persons. even, are believed to become dangerous when they decompose, and to cause serious diseases. Microscopic organisms multiply in them with great rapidity, and are disseminated in the surrounding atmosphere. For this reason, it is desirable that such matters should be removed from the vicinity of dwellings as quickly as possible. When they are discharged into sewers, their decomposition produces various gases— some of them very offensive—which are popularly known as sewer-gas, but should more properly be called, collectively, sewer-air. The most dangerous thing about sewer-air, however, is not the offensive gases, but the little organisms that float out with it into the streets or houses.

to the risk necessary for the proper care of the sick. If it becomes the duty of any other person to enter a sick-room, he should have in mind the following points: that the nose, on account of its narrow passages and extensive moist surface of mucous membrane, acts as a sort of filter, so that many impurities of the air are detained there and never reach the lungs, whereas, through the mouth, there is a straight and almost unimpeded course to those organs; that the body is less able to resist injurious influences of every kind when it is fatigued or in want of a fresh supply of food; and that matters escaping from the bodies of the sick and floating in the air are likely to settle on articles' standing in the room. Hence we deduce the following rules:

Never enter a sick-room when you are hungry or tired.

Always keep your mouth shut, except when talking.

Never eat or drink anything that has been standing in the sick-room.

CHAPTER VIII.

ASPHYXIA.

185. Asphyxia.—When the blood is deprived of its constant fresh supply of oxygen, the carbon dioxide produced in the tissues accumulates very rapidly, and in a short time the blood is brought into a condition in which it can not circulate, producing *asphyxia* or *suffocation*. The blood throughout the whole body then becomes venous. The arteries* as well as the veins are filled with black, sluggishly-moving blood. This black blood shows through the skin, particularly where it is very thin, as in the lips; and parts of the body which are usually of a healthy red or pink color become blue and livid. This blueness of the lips and of the flesh under the finger-nails is, therefore, a sure indication that the person is suffering from a lack of oxygen, and the only thing to do to save life, under such circumstances, is to supply fresh air. In drowning, strangling, poisoning by coal-gas or illuminating gas, this is always the great thing to be aimed at, and, as long as the heart beats, life exists, and consciousness can usually be restored.

* The arteries are those blood-vessels that carry the bright scarlet blood, which has received a fresh supply of oxygen, and the veins convey the dark blood, called venous, which is loaded with carbon dioxide and other waste matters, as hereafter explained.

186. Drowning. — The length of time during which a human being may remain under water and still recover, under proper treatment, is not yet actually determined. Young persons, it is known, live longer when submerged than older ones. As a rule, however, a person who has been entirely submerged for five minutes is dead beyond the possibility of resuscitation. And yet, even in such cases, attempts should be made, for any case may be an exceptional one.*

187. Resuscitation of the Drowned. — What, then, are the indications for the treatment of a person who is almost dead from drowning?

In the *first* place, he has been for some time deprived of oxygen. It is this which has made him unconscious.

In the *second* place, he has, probably, in his frantic efforts to breathe, taken water into his lungs, where it stops up the bronchi and air-vesicles, and must be cleared out before any air can enter.

In the *third* place, he is cold, and warmth, of itself, will do much toward bringing about his recovery.

In the *fourth* place, his circulation is at a very low ebb. The blood is so charged with carbon dioxide that it is sluggish, and, possibly, has almost ceased to flow.

We must *first*, then, *turn the person on his face*, and raise the lower part of the body somewhat, so as to let what water there may be in the lungs run

---

* Unconsciousness sometimes persists for a long time after a person has been removed into fresh air, when no special attempts at resuscitation have been made. It is said that persons have been restored by artificial respiration after they have lain unconscious and apparently dead for *five* hours.

out by the force of gravity. This action need occupy only an instant, for, if there be any water there,
it will immediately run out.

The person should then be laid *flat upon the back*,
without having the head raised, for we want the
first fresh blood to run to the brain, and the heart
is acting so feebly that it will be unable to send it
there if it has to propel it up-hill. The shoulders
should be raised a little by a pillow, a folded coat,
or other padding. All the *clothing should be loosened*
about the neck, chest, and waist, so as not to interfere at all with the movements of respiration.

The *wet, clinging clothing, if convenient, should be
removed entirely*, as it tends to keep up the chilliness
of the body. In any event, some one should attend
to the duty of *warming the body*, by rubbing it with
warm flannels, by bottles of hot water to the feet,
etc., etc.

In addition to these things, and chief of all, *artificial respiration* should be kept up until the patient
breathes naturally, or until absolutely all hope is
lost.

188. Artificial Respiration.—As the person lies
upon the back, the arms are to be grasped above
the elbows and brought upward above the head, so
as to touch, or nearly so. The large muscles of the
shoulder are attached to the walls of the chest in
such a manner that this movement of the arms raises
the ribs, and expands the cavity of the chest in very
much the same way that ordinary respiration does.
The chest being thus expanded, of course air rushes
in, and inspiration is effected. The arms should
now be returned to the sides of the body and pressed
against the ribs, when the chest-walls will recover

their former position by virtue of their elasticity, and expel all the air which had been taken in. This, it will be observed, is exactly the process of *natural expiration.* The rapidity of these movements should approach as nearly as possible to the rapidity of natural respiration—i. e., about *sixteen* or *eighteen movements to the minute,* and the drawing up of the arms above the head should occupy the usual time of inspiration. This process should be continued for hours, if necessary, and the first sign of recovery will usually be a slight change in the color of the lips and finger-nails to red or pink, indicating that the circulation and oxygenation of the blood have begun to be more active.

189. Additional Precautions.—During the whole process of resuscitation of a drowned person, care should be taken to keep the *mouth and throat clear of mucus and froth* by means of a finger covered with a towel. *The tongue must also be watched.* In persons who are almost dead and have lost their muscular power, this organ often slips backward into the throat, and covers the glottis so that no air can pass in or out. It is necessary, in such cases, for some person to take hold of the tip of the tongue with a towel to prevent its slipping from the grasp, and draw it forward so as to leave the passage to the lungs clear.*

As soon as the person begins to breathe he can swallow, and five grains of carbonate of ammonia should be given him in a quarter of a tumbler of

---

* In all cases of asphyxia, pure air is of the utmost importance. The sufferer should therefore be in a well-aired room, and whether indoors or out should never be surrounded by a crowd of people, whose respiration will pollute the air before it reaches the one who needs it most.

water, dry clothing should be placed upon him, and he should be put in a warm bed until his recovery is complete.

The above directions apply to all cases of suffocation, where there is no other injury to complicate the results of the mere deprivation of air.

CHAPTER IX.

THE HEART.

190. General Plan of the Circulation.—The *circulation* of the blood is brought about by a compli-

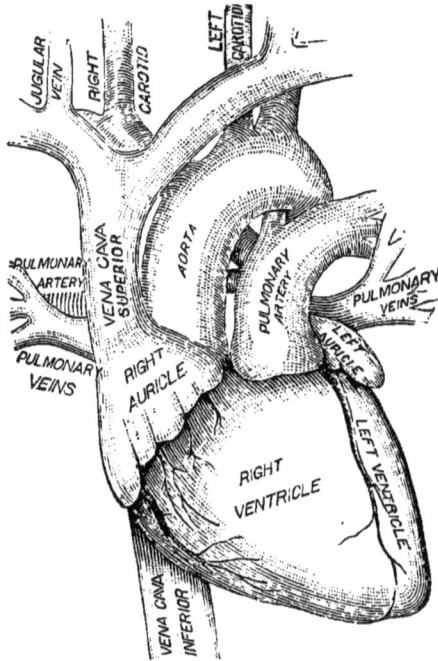

FIG. 41.—The heart and the large blood-vessels connected with it. The greater part of the left ventricle is hidden by the right ventricle.

cated series of tubes and channels, extending through every portion of the body, and all communicating with each other and with a powerful muscular central organ called the *heart*. The tubes are called, according to their structure, size, and function, *arteries, veins*, or *capillaries.**

191. The Heart.—The *heart* (Fig. 41) is a strong, hollow, muscular organ, lying behind the breastbone, with its greater portion to the left of it. It is shaped somewhat like a cone, with both ends rounded, and the larger end directed upward and toward the right. The lower end, or apex, is free to move in any direction, not being attached to anything, while the upper and larger end is held in place by the large blood-vessels which are connected with it and also with the spinal column. The whole organ is covered with serous membrane called the *pericardium*, and lies in a cavity which is also lined with serous membrane, so that, like the lungs and abdominal organs, its constant movements can go on with the slightest amount of friction.

192. The Double Circulation.—In order to understand the action of the heart, it is necessary to know, *first*, that there is a *double circulation* going on in the body at the same time. At every contraction of the heart a portion of the blood is thrown into the lungs and another portion into the remainder of the body; and these two portions never mingle with each other. To be more precise, and follow a particular mass of blood in its course through the body, we may state it thus: The blood starts, we will say, from a certain part of the heart; it goes directly to the lungs; thence it returns to the heart, but to a different part of the organ; then it goes out of the heart in the arteries to what is

---

* See Frontispiece.

called the general circulation—i. e., to all parts of the body, excepting the lungs; thence it is collected by the veins, and returns to the heart; at the next contraction it goes to the lungs again, and begins the same process; so that in this way all the blood passes through the lungs, and all of the blood visits all parts of the body; but in doing this it visits and passes through the heart twice. In short, it flows— 1. *From the heart to the lungs;* 2. *Back to the heart;* 3. *To the rest of the body;* 4. *Back to the heart.* Thus, there are two systems of circulation: one, called *the pulmonary circulation*, from the heart to the lungs and back again; the other, *the general circulation*, from the heart to the body and back again.

193. The Two Sides of the Heart.—This double and simultaneous circulation can not be brought about by a heart containing but one cavity. And, accordingly, we find that the heart is divided by a muscular partition, running lengthwise of the organ from front to rear, into two parts of nearly equal size, called the right and left sides of the heart. The *right side* carries on the *pulmonary circulation*, and the *left* the *general circulation*. So that the course of the blood is as follows: *From the right side of the heart to the lungs; thence to the left side of the heart; thence to all parts of the body; thence back to the right side of the heart.* If this order of the circulation be carefully observed, it will be seen that the *right side* of the heart never contains anything but dark or *venous blood*, and the *left side* always contains bright or *arterial blood*.

194. The Auricles and Ventricles.—Each side of the heart is divided into two cavities, making four in the whole organ. These cavities are called the

auricles and *ventricles.* The *ventricles* constitute the
greater part of the heart, and it is in their walls that
the greatest muscular power is located. The *auricles* are smaller cavities, situated at the upper extremity of the organ, and their walls are much
thinner and weaker than the walls of the ventricles.
The blood passes from the veins into the auricles,
from the auricles into the ventricles, and from the
ventricles it is forced out into the body. The course
of the blood, then, is from the body in general
*through the veins to the right auricle; from the right
auricle to the right ventricle; from the right ventricle to
the lungs; from the
lungs to the left auri-
cle; from the left au-
ricle to the left ventri-
cle; from the left ven-
tricle out to the body
in general,* whence it
is collected by the
veins and brought
back to the right au-
ricle, to begin the
same course again
(Fig. 42).

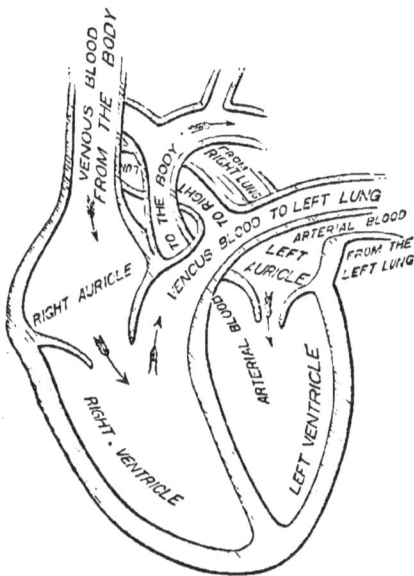

FIG. 42.—Diagram illustrating the course
of the blood through the heart.

**195. The Valves
of the Heart. —** At
the mouths of the
veins, where they
empty into the au-
ricles, there are no
valves, and they are not really needed at this point,
for the auricles do not contract with much force,
and as there is always a current in the veins running

11

toward the heart, and as the ventricles lie below the auricles, the blood naturally flows into the ventricles, where it meets with no resistance, rather than backward, where it would meet with considerable, having to oppose the force of gravity and also the current in the veins. In this manner the ventricles become filled with blood, and, when they begin to contract, the case is very different. Here there is an enormous pressure to overcome. The right ventricle must contract with force sufficient to send its contents into the lungs, pushing before it the column of blood already in the vessels. The left ventricle has to contract with a force sufficient to send its contents to the remotest parts of the body, also pushing along the blood which is already in the vessels. On the other hand, the resistance backward toward the veins is not nearly as great. The current of blood in the veins is not strong, and, even supposing that the resistance were equal in both directions, it is plain that the circulation would soon come to an end. The ventricle in contracting would force blood backward into the auricles and veins, and forward into the arteries, and then, when the heart relaxed, the blood would flow back again into the ventricles from both directions. This danger is averted by the introduction of *four sets of valves,* one between each auricle and ventricle, and one at the opening from the ventricle into the artery, through which the blood passes during contraction. The valves of the heart are double folds of the serous membrane which lines all the cavities of the organ, and are stiffened somewhat by a few fibers which run between the folds. All of these valves have three flaps, excepting the one

which separates the left auricle from the left ventricle, and this has only two.

The valves are all so constituted as to allow the
blood to pass only in one direction. The valves between the auricles and ventricles will allow blood
to pass from the auricles into the ventricles, but not
from the ventricles back into the auricles ; and the
valves at the mouths of the arteries will allow blood
to pass from the ventricles into the arteries, but not
from the arteries back into the ventricles.

196. The Blood-Vessels connected with the Heart.
—The large veins, by which all the blood from the
general circulation is poured into the right auricle,
are called the *venæ cavæ* (i. e., the hollow veins) ; the
large artery, by which the blood passes from the
right ventricle to the lungs, is the *pulmonary artery ;*
the large veins, by which the blood returns from
the lungs and enters the left auricle, are the *pulmonary veins ;* and the large artery, by which the blood
goes out from the left ventricle to all parts of the
body, is called the *aorta.*

197. The Circulation of the Blood.—The blood,
then, coming from all parts of the body in the veins,
enters through the venæ cavæ into the right auricle ; when the auricle is filled, it contracts and sends
the blood downward into the right ventricle ; when
the ventricle is filled, its walls contract, and the
blood passes into the pulmonary artery, its return
into the auricle being prevented by the closure of
the valves between the auricle and ventricle ; the
blood then goes through the lungs, and becomes
changed into arterial blood ; it returns to the heart,
to the left auricle, and passes from there into the
left ventricle ; the contraction of the ventricle then

forces it into the aorta, its return into the auricle being prevented by the valves; from the aorta it goes to all parts of the body, to be returned by the veins to the right side of the heart. The valves at the mouth of the pulmonary artery and the aorta prevent the blood which has entered them during the heart's contraction from flowing back into the cavity of the ventricle when it becomes relaxed.

198. Peculiar Valves in the Heart.—There is one peculiarity connected with the working of certain valves in the heart which is one of the most beautiful examples of adaptation in the whole body. It has been shown that there are no valves at the points where the veins enter the auricles, and still, when the auricles contract, the blood is not forced backward in the veins to any great extent, but passes downward into the ventricles. Some of the reasons for this have already been mentioned, but there is the additional fact that the opening between the auricle and ventricle on each side is very large, almost as wide as the auricle itself. There is, therefore, very little resistance, hardly any in fact, to the stream of blood passing from either auricle to the ventricle. But this large size of the opening might give rise to imperfect closure of the valves. The valves are made of thin sheets of membrane, stiffened a little by fibrous threads, but still very flexible. In the pulmonary artery and the aorta, the openings from the ventricles are so small that the valves are stiff enough to resist the backward pressure of the blood and keep the openings closed. The openings from the auricles to the ventricles, however, are so large that, if there were no special provision to prevent it, the valves would not only be pressed back-

ward toward the auricle when the ventricle contracted, so as to meet at their edges and close the
opening, but, on account of their flexibility, their
borders would be bent still farther back, so as to
open into the auricle, and allow a reflux of blood
into that cavity.

This difficulty is obviated in the following manner : There are numerous fine but strong fibrous
threads or cords attached to the edges of the valves,
and from that point running downward to the walls
of the ventricle. These cords are just long enough
to allow the valves to close perfectly, but not pass
any farther back toward the auricle. But here another difficulty arises. If the cords are long enough
to allow the valves to close at the beginning of the
contraction of the ventricle when the cavity is at
its full size, then they will be too long when the
contraction is toward its end and the cavity is diminished in size, and allow the valves to be pressed

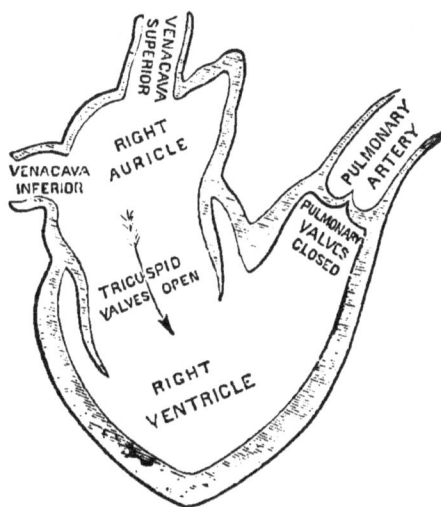

FIG. 43.—Illustrating the action of the valves in the right side of the heart.

too far back.   In other words, to fulfill their object,
these cords must be able to become longer or short-

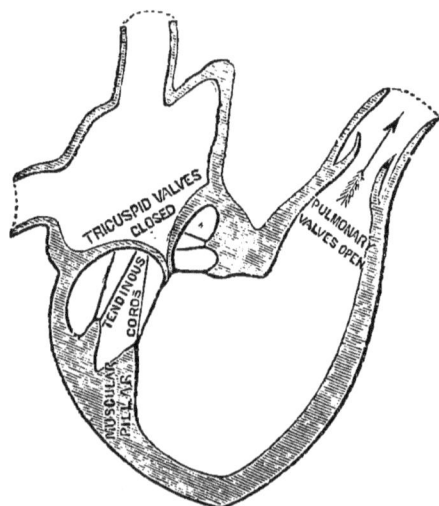

Fig. 44.—Illustrating the action of the valves in the right side of the heart.

er according to circumstances.   This is effected by
small muscular projections or pillars, which extend
from the walls of each ventricle into its interior, and
to which the cords are attached.   When the heart
contracts, these little pillars contract at the same
time, and make the cords attached to the valves
shorter and shorter as the contraction proceeds, just
in proportion as the cavity of the ventricle grows
smaller and smaller.   In this way the reflux of blood
through these large valves is prevented (Figs. 43,
44).

199. The Contraction of the Heart.—The contraction of the heart does not run successively from
one auricle to the corresponding ventricle, and then
from the other auricle to the other ventricle, but the
contraction of both sides of the heart is simultane-

ous. It begins at the auricles and extends downward until the ventricles are both firm and hard and
reduced to their smallest possible size. The organ
then becomes relaxed, and is for an exceedingly
short time quiet. During this stage of relaxation
the auricles are being filled with blood from the
veins, and there is also a current running into the
ventricles from the auricles. During the stage of
contraction the blood is being forced into the circulation through the aorta and the pulmonary artery.

200. The Sounds of the Heart.—These alternate
contractions and relaxations of the heart are accompanied by *sounds*, which are very audible to any one
who applies his ear to the region of the heart in a
living person. These sounds are two in number, the
first being a prolonged, rumbling sound, and the
second short and sharp. The *first* sound is made
during the time when the heart is contracting, and
the *second* just at the end of the contraction, or beginning of relaxation. The *first* sound is supposed
to be produced partly by the closing of the large
valves between the auricles and ventricles, which
occurs just at the moment when this sound begins,
and partly by the contraction of the muscular fibers
of the heart. The *second* sound is positively known
to be produced by the closing of the pulmonary and
aortic valves. It is by the variation in distinctness
and quality of these sounds, and the addition of
other sounds to them, that physicians are enabled to
determine with wonderful accuracy the condition of
the valves of the heart.

201. Nervous Supply of the Heart.—The heart
is plentifully supplied with nerves, which regulate
its movements. One set, belonging to the cerebro-

spinal system (which will be treated in the chapter on the nervous system), have the power of checking or arresting the heart's action, and are therefore called the inhibitory nerves of the heart. Another set, having an opposite function—i. e., that of increasing the heart's action—are called the accelerator nerves. If the inhibitory nerve is destroyed, or temporarily paralyzed, the pulsations of the heart are suddenly increased. If the accelerator nerve is paralyzed, the heart ceases to beat.

202. **Rapidity of Pulsation in Health.**—The *contractions* of the heart take place with regularity, and average in the adult about *seventy per minute.* The rate is higher in children and women than in men, and this fact is probably connected with their greater impressionability. The heart-pulsations appear to be slower in proportion as the individual is cool and deliberate in his judgments. The pulse of Napoleon Bonaparte is said to have averaged only forty-four to the minute, and is one of the slowest on record. Sudden emotions may increase its rapidity and force, so that a process of which we are usually unconscious becomes very perceptible and unpleasant, or, on the other hand, they may cause it to stop for a moment altogether, to skip a beat, as it were, producing the sensation of "fluttering" at the heart. Although the action of the heart is thus influenced by our feelings, it is beyond our control. Its pulsations are ceaseless and regular, until interrupted by disease or death. But, notwithstanding this general fact, there are some instances on record of persons who have been able to affect the action of the heart by an effort of the will. The most remarkable one of these, perhaps, was a

Colonel Townsend, of Dublin. This person, on several occasions, in the presence of medical men, lay down and caused the contractions of his heart to become so faint as to be imperceptible. During the experiment the circulation was so far interfered with that he became pallid and unconscious. After a half-hour or so, he would gradually return to his natural condition. As might have been expected, he performed the experiment once too often. He stopped the action of the heart for the last time in the same way as he had done before, and it never resumed its work.

203. Effect of Alcohol upon the Heart.—A small quantity of alcohol increases the frequency of the heart's beats by paralyzing for a time its inhibitory nerve. The effect may be compared to removing the balance-wheel from a watch or the drag from a wagon-wheel as it is going down-hill. The proper restraint being taken off, the heart's movements become at once more hurried and frequent. It is a wise provision of Nature that the accelerator nerves belong to a part of the nervous system that is much more slowly influenced by any disturbing agent than the inhibitory. If this were not the case, and the accelerator nerve were as quickly affected as the inhibitory, life would be in frequent jeopardy from the effects of injurious substances which men either recklessly or ignorantly take into their systems. Whenever the heart is compelled to more rapid contraction than is natural, it has less time to rest. Although this organ seems to be constantly at work, it really rests more than half the time, the time occupied by a contraction being to the time between the contractions about as two to three, so that, although the periods of relaxation are very short, they are so numerous (seventy or more a minute) that the aggregate amount of rest in a day is very great. Now, if the rapidity of the contractions is increased materially and continuously, although the aggregate amount of time for rest may be the same as before, yet the waste caused by the contractions is greater, while the time for rest after each one is shorter. This lack of rest produces exhaustion of the heart-muscle, ending in partial change of the muscular tissue into fat. The heart then becomes flabby and weak and its walls become thinner, a condition known to physicians as a "fatty heart," often resulting in sudden death.

204. Effect of Tobacco on the Heart.—The use of tobacco often produces functional derangement of the heart's action. The pulse becomes weak and intermittent, and palpitation of the heart is common. These effects are so well known that symptoms of this kind in a boy or man, without any discoverable organic disease to account for them, are at once put down by physicians as due to a "tobacco heart." This weakness of the heart is probably due to the influence of tobacco upon its nervous supply, and its immediate effect is seen in a languid circulation, imperfect blood-changes, pallor of the face, and occasional fainting-spells, especially in the young. These are the symptoms oftenest produced by the smoking of cigarettes, and they have been sometimes attributed to the paper in which the tobacco is wrapped, or to the drugs with which it was adulterated. But this is mere beating about the bush. The most hurtful ingredient of the cigarette is the tobacco, just as truly as the most hurtful ingredient of fermented drinks is the alcohol.

CHAPTER X.

205. The Blood-Vessels.—The heart, although a very powerful organ, would not be able to force the blood through the whole body, and back to itself again, without assistance, and this assistance is furnished by the structure of the blood-vessels themselves. The blood leaves the heart by the arteries and comes back to it through the veins, and these two systems of vessels differ very much in their structure.

206. Structure of the Arteries.—The *arteries* are tubes, with strong walls, described by anatomists as having *three layers.* The *innermost* is a delicate, smooth membrane ; the *middle* one is composed of elastic fibers and also fibers of the non-striated or involuntary muscular tissue ; the *outer* one is made up of strong connective tissue. Thus the walls of the arteries are very.elastic, and, if the tube is distended, it returns to its former size as soon as the internal pressure is removed.

207. The Pulse.—When the heart contracts, its contents are driven with great force into the arteries, and, as the blood already contained there resists somewhat the advance of the fresh supply, the walls of the arteries are stretched to accommodate the mass of blood which is thrown into them. When the heart relaxes, and the pressure from that direction is removed, the elastic walls of the arteries react upon their contents, and, if it were not for the valves, would drive the blood, or a portion of it, back into the heart. At the slightest backward pressure, however, the valves close, and the elasticity of the arteries thus gives the blood another impulse forward toward the surface of the body. The impulse given by the heart's contraction, together with that caused by the recovery of their natural position by the walls of the arteries, gives rise to the *pulse*, which can be felt at any point in the body where an artery runs near enough to the surface. The common place of feeling for it is in the wrist, merely because that is the most convenient and accessible; but it may also be felt in the ankle, in the neck, in the temple, or in the upper arm.

208. The Capillary Blood-Vessels ; their Structure.—The large vessels, by which the blood leaves the heart, viz., the pulmonary artery and the aorta, divide and subdivide continually, the branches growing smaller and smaller as they approach their termination, their walls at the same time undergoing a change in structure. The elastic tissue, which is so abundant in the larger arteries, gradually disappears as the vessels diminish in size, and the muscular tissue becomes more prominent, until even this finally vanishes, and the smallest blood-vessels, called the capillaries, are composed of a thin membrane, not divisible into layers. Thus the largest arteries are very strong and very elastic, while the smaller ones lose in elasticity, but, from the amount of muscular tissue they contain, are very contractile.

209. Size of the Capillaries.—The *capillaries*, in which the arteries finally end, are only about the $\frac{1}{3000}$ of an inch in diameter—just large enough to allow the blood-corpuscles to pass through them, so to speak, in single file. Their number is beyond computation. They are so thickly strewed in the body that the point of a fine cambric needle can not anywhere be inserted between them. As every one knows, it is impossible to find an instrument with a point so fine as not to wound a blood-vessel if introduced through the skin. These vessels are entirely indistinguishable to the naked eye, and, before the discovery of the microscope, it was a great problem for anatomists to explain how the blood got from the arteries into the veins, as they could find no direct communication.

210. The Veins.—After passing through the capillaries, the blood enters the *veins*. These vessels contain in their walls much less muscular and elastic tissue than the arteries, and more connective tissue. The consequence of this is, that the walls of the veins are flaccid and yielding, and, if they are cut across, the sides fall together and tend to close the opening. If an artery, on the other hand, be cut, the tube remains open and in a sense rigid, although, as will soon be shown, its caliber is somewhat diminished. The veins, very minute at first, gradually unite and become larger and larger, until finally all the veins of the general circulation form two large vessels, called the *venæ cavæ*, which discharge their contents into the right side of the heart—one vena cava receiving all the blood from the head and upper extremities, and the other that from the rest of the body.

211. Circulation of Blood in the Veins ; Influence of Respiration.—The *circulation* of the *blood* in the *veins* is brought about in *three ways :* In the *first* place, the act of respiration has its influence. When the chest is expanded by muscular action, every fluid which is outside of it tends to rush in and fill the enlarged cavity. The chief space is filled by the air, as that is more perfectly fluid and meets with the least resistance from friction. But the blood is also drawn in through the veins, and the real extent and power of this suction can be very easily seen whenever the entrance of air is impeded. In such cases the veins in the neck can be plainly seen to become swollen and full during expiration, and emptied again during inspiration.

212. Influence of Muscular Contraction.—In the *second* place, the contraction of the voluntary muscles aids in the return of the blood to the heart. While the arteries, as a rule, run deep in the body, out of the reach of injury, the veins are largely near the surface, and the whole exterior of the body is more or less streaked by the blue lines which indicate their course. Now, during the contraction of a muscle, it not only shortens but becomes broader and thicker, and, of course, compresses to a greater or less degree everything near it. Thus the veins are continually being pressed upon here and there, in various parts of the body, during the whole of our waking hours, and even to some extent during sleep.

213. The Valves of the Veins.—But merely pressing the blood out of a certain portion of a

vein might send it in either direction; it would be
almost as likely to send it away from the heart as
toward it. This reflux of blood in the veins is prevented by *valves* (Fig. 45), which allow the blood to
pass through them readily toward the heart, but
not away from it.* These valves are particularly

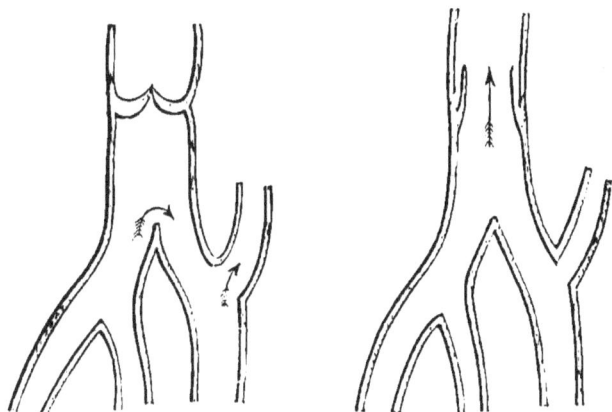

Fig. 45.—Diagrams illustrating the action of the valves in the veins.

numerous in the lower extremities, for here the
force of gravity acts in opposition to the current of
blood, and would seriously interfere with the circulation if there were no special provision with reference to it.

Thus, when blood has been forced out of a portion of a vein by pressure, it can not go backward
on account of the valves, but must go forward in
every case. This fact and the action of the valves
may be beautifully seen in the arm of any person,
where the veins are not obscured by too much fat
beneath the skin. If a place be chosen where a vein

* It is said that the discovery of the proper working of these valves
first suggested to William Harvey the true theory of the circulation of
the blood.

is visible, with no branches for an inch or so, and one finger be placed upon it so as to stop the flow of blood, the portion of the vein on the farther side from the heart will be seen to fill with blood, and at some point will probably look swollen. This slight swelling marks the situation of a valve. If a finger be passed along a vein toward the heart, pressing upon it all the time, the vein will be seen to fill behind the finger; while if the finger be passed in the opposite direction, away from the heart, the vein will be empty and collapsed behind the finger, and perhaps hardly noticeable. This clearly indicates the direction of the current of blood.

214. Influence of the Pressure in the Capillaries. —But the *third* cause of the venous circulation, and the most important of all, is the blood which is constantly accumulating in the capillaries and exercising pressure on the column of blood already in the veins. This pressure is unceasing and powerful.

These three causes, acting together, keep up a free and steady flow of blood in the veins toward the heart.

215. Communicating Blood-Vessels.—In both arteries and veins, there are numerous communicating branches, so that, when a blood-vessel is obstructed, the blood passes out into other vessels and around the point of stoppage, and, excepting in extraordinary cases, the nutrition of the part is not interfered with.

216. Recapitulation : Rapidity of the Blood=Current in the Vessels.—The arteries, then, carry the bright scarlet, highly oxygenated blood from the heart out to all parts of the body for its nutrition.

It is sent to the remotest capillaries, partly by the contraction of the heart, and partly by the elasticity of the arteries; from the arteries it enters the capillaries, where the essential but very obscure processes of nutrition are carried on. It has been found that the current of blood rushes through the arteries with an average velocity of twelve inches per second, but, in consequence of the smallness of the capillaries and their distance from the heart, as well as the magnitude of their combined areas as compared with that of the aorta, the blood moves through them very slowly, not faster, it is thought, than one thirtieth of an inch per second. When the capillary circulation is looked at through a microscope, as it may be in the web of a frog's foot,* it is seen that the red corpuscles pass along through the minute vessels, sometimes two together, but oftener in single file, and without much trouble; but the white corpuscles are more affected by friction, and drag along, sticking fast here and there until they are started again by the current. During the passage of the blood through the capillaries, certain of its ingredients transude through the walls of the blood-vessels, and lie in immediate contact with the tissues outside; these are the nutritive materials by which the various tissues of the body are kept in repair; the cells select their nourishment, and what is left, together with waste and used-up matters from the cells, is taken up by the lymphatic vessels and returned to the large

* The foot of a live frog may easily be fastened by strings so that it can be placed under the microscope. The thin membrane is transparent, and the circulation of the blood, as seen in this way, is perhaps the most surprising and instructive sight that can be witnessed.

veins near the heart. These matters constitute
what is called the lymph. Changes in the gaseous
constituents of the blood also take place in this part
of the circulation, and so we find that, when the
blood emerges from the capillaries into the veins,
it has become of a dark-purple color, and unfit for
further use in the body until refreshed. So the
process which takes place in the capillaries is in
some degree the reverse of that which takes place
in the lungs. The blood enters the lungs of a black
or deep-purple color and comes out bright red. It
enters the capillaries bright red and comes out dark
purple. It then passes back to the heart through
the veins, the steady flow being maintained partly
by the suction caused in the act of respiration, partly by muscular contraction and consequent pressure
on the veins, and mainly by the pressure from the
capillaries, which constantly forces the blood on-
ward.

217. Peculiarity of Pulmonary Artery and Veins.
—There is one exception to the rule that the arteries
carry scarlet blood, and one to the rule that the
veins carry purple blood. The *pulmonary artery*
carries *venous blood* from the right side of the heart
to the lungs, and the *pulmonary veins* bring back
scarlet or *arterial blood* from the lungs to the left
side of the heart.

218. Rapidity of Venous Circulation.—The rapidity of the current in the veins is estimated at
about two thirds of that in the arteries, or about
eight inches per second. As all the blood which
goes out through the arteries must return through
the veins, it might be inferred that the velocity of the
flow in both systems of vessels would be the same.

If the capacity of the vessels were the same, it would necessarily be so, but there are generally, with rare and unimportant exceptions, two veins returning the blood sent out by one artery, so that the capacity of the venous system is as a whole about twice that of the arterial, and the velocity would be half as great; but, if we take into account the difference in the distention and fullness of the two systems, the estimate given above is probably nearly correct.

219. Rapidity of the General Circulation.—An interesting question arises with regard to the *rapidity* of the *general circulation.* Experiments have been made which show that this is somewhat greater than would have been expected. A substance which remains unaltered in the blood, and which can easily be detected by chemical means, was introduced into a large vein on the right side of the neck. It was plainly detected in the blood drawn from the vein of the left side, in from twenty to twenty-five seconds. In this short time, the blood in which the substance was introduced must have gone down to the right side of the heart, from there to the lungs, thence to the left side of the heart, and thence through arteries to the head, before it entered the vein in which it was detected on its way back to the heart.

The time required for all of the blood in the body to pass through the heart can not be accurately determined, but only estimated. It probably varies very much with the vigor of the heart's action, the amount of exercise taken, the frequency of the respiration, etc. In the dead body, however, each ventricle is found to contain about two ounces of fluid. Now, one ventricle is to be estimated in the calculation, because, in order to complete the entire round of the body, all the blood must pass through the left ventricle. If two ounces enter and leave the ventricle at every contraction of the heart, and there are seventy pulsations in a minute, one hundred and forty ounces, or eight and three quarter pounds, will pass through the organ in this short time. Estimating, then, thirteen pounds of blood as the average amount in an adult, two minutes at the most would suffice for the completion of the circulation, and this is probably pretty near the truth.

220. The Supply of Blood in any Part varies.— The amount of blood in any portion of the body at any particular time depends upon certain relations which exist between the blood-vessels and the nervous system. The walls of the arteries are plentifully supplied with involuntary muscular fibers. The contraction of these fibers diminishes the caliber of the artery. They are most abundant in the small arteries, and their contraction or relaxation is controlled by certain nerves called *vaso-motor nerves*, because they control or cause motion in the vessels to which they are distributed. If the nervous stimulus be such as to cause a contraction of the arteries supplying any particular part of the body, the supply of blood to that part will be diminished, and will be diminished in exact proportion to the amount of contraction in the blood-vessels. If, on the other hand, the nervous control be altogether withdrawn, and the arterial walls completely relaxed, the amount of blood in the part affected will be increased to a corresponding extent.

221. Effect of Alcohol on the Blood-Vessels.— Alcohol, even in a moderate dose, produces dilatation of the superficial blood-vessels, by narcotizing or paralyzing the vaso-motor nerves. This accounts for the flushing of the face which follows so soon after the drinking of any alcoholic liquor. This dilatation of the small vessels, which produces such visible effects on the external surface of the body, also takes place in parts hidden from sight. The brain, lungs, heart, liver, kidneys, and stomach are all congested, and the secretions increased in amount.

222. Diseases of Blood-Vessels caused by Alcohol.—If alcoholic drinks are used habitually, this dilatation becomes permanent, partly from their effect on the vaso-motor nerves, and partly from a loss of elasticity in the walls of the vessels themselves, owing to a deposit of a fatty nature, which occurs in spots between the middle and internal coats, making the walls of the vessels weaker at such points. The pressure of the blood produces a gradual stretching and thinning of the wall of the vessel at these weak spots, resulting in the formation of a hollow tumor, like a bladder, on the side of the vessel, in which the blood eddies round, like the water in a whirlpool. This disease is called aneurism, and is a very distressing and dangerous one. The tumor presses upon neighboring nerves, causing obstinate neuralgias and sometimes paralysis, and when the outside coat has become so thin that it can resist the pressure of the blood no longer, it bursts, and a fatal hæmorrhage is the result. The loss of elasticity in the arteries, produced in the way just described, also disastrously affects the circulation, for when they are filled with blood they do not react sufficiently upon it. When the heart contracts again, therefore, and throws a fresh supply of blood into them, it has to be forced along in large part by the heart alone, whereas in healthy persons the arteries do a good share of the work. Thus the heart becomes enlarged and contracts more powerfully at the same time that the walls of the vessels are weakened by disease, and they often burst under the increased pressure, causing what is known as apoplexy.

223. Features of the Habitual Alcohol-Drinker.—The permanent dilatation of the surface blood-vessels produces a permanent flushing of the face in drinking men, and often some of the veins are so enlarged that they show as red streaks upon the cheeks and nose. The white of the eye, for the same reason, becomes red. The circulation of the blood in these dilated vessels being sluggish, the nutrition of the skin is interfered with, and it may be covered with red blotches, or disfigured by pimples, especially on the nose. The circulation in the latter organ being naturally more feeble than elsewhere on the face, it shows these effects earlier and more plainly, and sometimes becomes almost purple in color and knobbed and swollen at the end.

224. The Aorta.—The blood destined for the general circulation all leaves the heart through the *aorta*. This is a large vessel, about five eighths of an inch in diameter, with thick, strong walls, as it has to bear an enormous pressure. It begins at the upper end of the left ventricle and on its right side, and, after leaving the heart, springs upward toward the right, near the breastbone, to the second rib; then arches backward, and passes between the lungs to the back; here it curves again, and runs down along the spine, through the diaphragm, to the lower portion of the abdomen—all the way lying in front of the vertebræ. When it reaches the pelvis it divides into two branches, one of which goes to each lower extremity.

225. The Femoral Arteries.— These branches run the same course in either limb. They emerge from the abdomen on the front of the limb, in the groin, not far beneath the skin, and pass in as straight a line as possible on the inside of the thigh to a point behind the knee-joint, at about the middle of the hollow space which is found there between the tendons on each side. In its course along the thigh the vessel is called the *femoral* artery (Fig. 46). At the knee it again divides into two branches, one of which runs down in front of the leg, and the other behind to the foot, where they further subdivide, to supply each of the toes with a small artery on each side. All through this course, however, arteries of different sizes are given off as branches to supply the different organs in the chest and abdomen, as well as the muscles and skin, the course of the main arteries only being here indicated.

FIG. 46.—The right thigh. The dotted line represents the course of the main artery (the femoral).

226. The Brachial Arteries.—From the arch of the aorta spring upward the vessels which supply the neck and head and upper extremities. Two large vessels go to the arms, one to each. They pass upward and outward between the collar-bone and the first rib, and dip down from the neck, passing to the arm through the arm-pit. As soon as the artery enters the arm it is called the *brachial*, and it con-tinues its course down the inside of the limb to the elbow, where it comes in front (Fig. 47). Here it divides in two, the *radial* and *ulnar*, which pass down, one on each side of the arm, to the hand, where their subdivision furnishes a small artery for each side of each finger and the thumb. These vessels also, throughout their course, give off branches to the muscles and other parts.

Fig. 47.—The left upper arm. The dotted line represents the course of the main artery (the brachial).

The *radial* artery is the one ordinarily felt in order to judge of the pulse, and is easily found on the radial or thumb side of the arm, about an inch above the fold where the hand bends upon the arm.

227. The Carotid and Vertebral Arteries.—Four arteries supply the head and face. These are the two *carotid* arteries in front and the two *vertebral* behind. The two *carotids* pass up on each side of the neck, and, when they approach the skull, divide into two main branches, one of which supplies the face, while the other enters the skull and supplies the brain. The *vertebrals* supply the brain, and pass up to it almost the entire distance inside the bones of the spinal column. It is important to know the course of the *carotid*. There is a powerful muscle in the neck, which passes upward from the upper end of the breastbone and parts in its vicinity to a point just behind the ear, where it is attached to the skull. Its contraction turns the head, or, with other muscles, bends it over to one side or the other. Its outline is distinctly perceptible under the skin, particularly when it is somewhat contracted, and the head thereby a little twisted. The *carotid* artery runs very nearly along the anterior border of this muscle, and its beating may be readily felt in this situation.

228. The Large Veins.—Each artery is usually accompanied by one or two veins—the largest by one, and the smaller ones by two. The veins are called, as a rule, by the same name as the corresponding artery, the most notable exception to this being the *jugular* veins, which are the companions of the *carotid* arteries, and run close by their side. All the blood from the lower part of the body is finally collected in a large vein, called the *vena cava inferior*, which runs up the spine beside the *aorta*, while that from the head, neck, and upper extremities is collected in the *vena cava superior*, and both these large veins discharge their contents into the right auricle of the heart. The veins which take the blood from the digestive organs unite to form the *portal vein*, which enters the liver, and the large vein which emerges from the liver joins the *vena cava inferior*, so that all the blood from the digestive organs must go through the liver before it enters the general circulation.

The *pulmonary artery*, soon after it leaves the right ventricle, divides into two branches, one of which passes under the arch of the *aorta*, before described, to the right lung, and the other goes to the left lung. The corresponding *pulmonary veins* are two from each lung, and they empty their contents into the left auricle.

229. Obstruction of the Circulation.—If a cord be tightly bound about a finger, that part of the member which is farthest from the heart will soon become livid and begin to swell. This effect is due to the difference in structure and position of the arteries and veins. The arteries are tubes with stiff, elastic walls, and do not usually lie very near the surface, whereas the veins have thin, inelastic walls, and many of them are very superficial. The consequence is that it is a somewhat difficult matter to compress an artery so as to entirely prevent the flow of blood through it, while the veins are very easily compressed. In binding a cord around the finger, then, unless great force be applied, the veins are compressed and the current of blood in them checked, while that in the arteries is not at all or only slightly interfered with, so that blood is being continually carried into the finger, and can not flow out. The accumulation of blood in the part accounts for the swelling, and the dark color is that of venous blood. If, now, the cord be removed, the swelling does not immediately disappear, because wherever there is a damming of the current of blood so that its flow in the veins is interfered with, after the vessels become distended to a certain point, the pressure on their walls is relieved by the loss of a portion of their contents. The serum of the blood begins to pass through the walls of the veins into the tissues outside of them, producing the condition called *dropsy.** When the circulation of the part is restored, all this serum which has left the vessels has to be reabsorbed, partly by the blood-vessels and partly by the lymphatics, and this occupies an appreciable time.

Now, let us suppose that this obstruction, instead of being applied to the veins on the surface of the body, is situated in some of the interior organs. It is plain that a similar effect will be produced if the supply of blood to any part remains the same, while the return of it is prevented. Such obstructions are most common in the heart and lungs, as a result of disease in those parts.

230. Disease of the Heart.—Let us suppose that the valves between the left auricle and left ventricle have been inflamed, and have become so altered in their shape and size that, when the heart contracts to force the blood from the left ventricle into the aorta, they do not completely close the opening. It is plain that a portion of the blood will be forced backward into the auricle. This regurgitation, as physicians call it, takes place at every beat of the heart, and as the auricle is in this way kept filled with blood, and a sort of conflict continually takes place between the current coming into it from the

---

* When fluid passes in this way from the interior of the blood-vessels through their walls into the tissues outside of them, the fibrin does not form a part of it. In other words, it is not the *plasma* that is effused, but the *serum.*

lungs and that coming back to it from the ventricle,
the current in the pulmonary veins is materially
interfered with. In short, the blood is dammed
backward into the lungs. Here, then, is an actual
obstruction to the circulation, as much as there is
when we tie a string around the finger. The result
can be easily ascertained by following backward
the course of the circulation. The obstruction to
the exit of blood from the lungs causes the blood
to accumulate in that organ—in other words, produces a congestion there. The blood, not being
changed frequently enough, does not get enough
oxygen, and the person is obliged to breathe faster
in order to supply more. This we call *shortness of
breath*. The accumulation of blood in the lungs,
in a measure, obstructs the current of blood which
comes to them through the pulmonary artery. The
blood in the pulmonary artery being hindered in its
course, the right ventricle is not able to empty itself perfectly, becomes dilated, and its walls become
thinned. The obstruction at this point prevents the
right auricle from emptying itself properly, and
this interferes with the free return-current in the
large veins which are connected with it. Thus the
flow of blood is hindered in the veins from all parts
of the body by the disease of one set of valves. The
obstruction of the venous flow brings about the
same results that we have observed in the constricted finger. The face becomes more or less
livid, the lips and finger-nails bluish, and the serum
of the blood passes out into the tissues around the
veins, causing *general dropsy*. This is the most common form of *heart-disease*, and it has been somewhat
minutely described, in order to show how purely

mechanical are some of the diseases to which we
are subject. There are other forms of disease of
the valves, whose effects may be traced by those
curious in such things. The aortic valves, for example, may be affected so as to close incompletely,
and allow a part of the contents of the aorta to be
forced back into the ventricle by the elasticity of
the artery every time the heart relaxes. On the
other hand, these same valves may be made so rigid
by disease that, although they close tightly enough
to keep the current of blood from setting back
through them, they do not open sufficiently wide
to allow a free flow through them during the heart's
contraction. This latter form of disease is one of
the most common, and gives rise to symptoms very
much like those of the disease which has been more
fully described above. Sometimes, as a result of
inflammation of a peculiar kind, the edges of the
valves have small, wart-like masses attached to them.
These little masses sometimes prevent the valves
from closing properly, and sometimes not, but they
always offer more or less obstruction to the circulation. Occasionally one of these bodies becomes
detached from the valve by the force of the blood-
current, and is whirled away through the body.
As the current of blood in the arteries is continually
in the same direction, and they grow smaller and
smaller, such a little body at length reaches a spot
where the caliber of the artery is too small to let it
through. It plugs up the artery. This gives rise to
different consequences, according to circumstances.
Sometimes the circulation is carried on, in spite of
the obstruction, by other arteries which pass around
the point of plugging ; and sometimes the part which

receives its supply of blood from the affected artery, being thus suddenly deprived of it, dies, falling into the condition known as *gangrene*.

231. How Heart-Disease is detected. — All of these affections of the valves of the heart, interfering as they do with the free flow of the blood, give rise to *sounds* of greater or less intensity, the varieties of which are familiar to practicing physicians, and indicate to them quite accurately the character and extent of the disease.

232. Effect of the Coagulation of the Blood.— Whenever the surface of the body is wounded, blood-vessels are necessarily severed, and, if there were no means of stopping the consequent escape of blood, it would not take very long for the whole body to be drained. The fibrin of the blood, however, by its property of coagulation, serves to arrest bleeding. All of the methods used by surgeons to stop hæmorrhage in any part of the body have for their object the coagulation of the blood.

233. Conditions of Coagulation.—Blood coagulates much more rapidly on a rough, ragged surface than on a smooth one, and in a wound that is much bruised or lacerated there is often very little bleeding, even from large arteries. Instances have been known in which the arm has been violently torn from the body by machinery, and the brachial artery divided, with comparatively little bleeding; although, if the arm were cut off, without controlling the main artery, the person would die of hæmorrhage in a few minutes. In a clean-cut wound the hæmorrhage is often very severe from small blood-vessels, and almost always needs some artificial control.

234. Arterial and Venous Hæmorrhage.—When an artery has been severed, it is not difficult for one who understands the circulation to detect it. In the *first* place, the arterial blood is of a bright scarlet color, and, in the *second* place, it comes from the vessel in jets or spurts. The blood in the veins is dark purple, and, as the veins do not pulsate like the arteries, it flows from the wound in a steady, uniform stream. The force with which the blood issues from a cut artery is surprising. A jet from one of the little arteries of a finger will spurt at least a foot in a stream no bigger than a knitting-needle. A vein never bleeds in spurts. In a wound, however, both veins and arteries may be severed and the blood mixed, although usually either the arterial or venous color predominates. If the blood from the wound be soaked up by a sponge or soft cloth, an arterial spurt can be seen before the cut fills up again with blood.

235. Natural Arrest of Hæmorrhage.—If a wound be left to itself, the following are the means provided by Nature to arrest bleeding: When an artery is cut, its walls always contract somewhat, so that its caliber is diminished. The elasticity of the artery also draws it backward to some extent into the flesh. In the smallest arteries, these acts are often sufficient to stop the bleeding entirely. In every case they resist the current, make it move more slowly, and so favor coagulation. When a vein is cut, the walls, being thin and inelastic, collapse, the opposite sides coming in contact and tending to obstruct the flow of blood. In these ways the current of blood from all cut vessels is diminished in force. The blood begins to coagulate almost immediately, and this offers still further resistance to the out-coming current. If the bleeding continues, and the blood-vessels are so drained that the brain feels the lack of blood, the person faints, the nervous force of the heart is diminished, and in this way also the force of the flow is lessened. This is Nature's last resort, and, if the vessels injured are so large that these means are not sufficient to stop the flow, the person will bleed to death.

236. Artificial Arrest of Hæmorrhage ; Cold.— The artificial means which we have at our command for arresting hæmorrhage merely aim to assist these attempts of Nature.

The application of cold to any part of the body produces pallor, caused by a diminution in the supply of blood to the part. This is owing to the fact that the stimulus of cold causes contraction of the smaller blood-vessels, and so lessens the amount of blood in them.* When a cut surface is exposed to the air, if the vessels which have been severed are small, the coldness of the air is sometimes sufficient to stop the bleeding by causing a contraction of the vessels. This effect can be increased by bathing the wound in cold water. If this be done, however, the cut surface should not be wiped or rubbed with the sponge or towel, for fear that the already coagulated blood, which begins to form an obstruction to the flow, may be washed away.

237. Styptics. —We also have artificial means of bringing on coagulation of the blood. The sub-

---

* The diminution in the caliber of arteries in such cases is caused by the contraction of the involuntary muscular fibers which surround them, as previously explained.

stances used for this purpose are called *styptics.* The *persulphate of iron* is an exceedingly powerful one. The *muriated tincture of iron* is another. *Alum*, or *tannic acid*, or any other astringent, produces the same effect. The great objection to all of these substances is, that they so alter the tissues with which they come in contact that the wound is often hindered from healing as rapidly as it otherwise would. They are mostly used when it is impossible to apply actual pressure upon the bleeding vessels.

238. Compression.—*Compression* is the most perfect and unobjectionable method of arresting hæmorrhage. This brings no outside injurious matter in contact with the wounded surface, and acts merely by stopping the flow temporarily until Nature has time to stop it permanently.

If the wound be small, and can be covered by the finger, it will generally be enough to put the finger or thumb immediately on the cut and press toward a bone until the bleeding stops. If the wound be too large for this, pressure must be made at a point outside of the wound, where the vessels which are supposed to be cut are known to run.* If the bleeding be venous, of course the current is running toward the heart, and the pressure must be made on the side farthest from that organ. If the hæmorrhage be arterial, the pressure must be made between the wound and the heart. In the

* Or some soft, porous material, such as a towel, handkerchief, sponges, or cotton cloth or batting, may be crowded into the wound and held there by the hand or a bandage. This is often the easiest way to apply pressure directly to the mouths of the bleeding vessels. Woolen material is too rough and stiff for this purpose.

case of venous hæmorrhage, a comparatively slight pressure will generally be sufficient, as the walls of the veins possess little resisting power, and are readily forced into contact. On the arteries, however, greater force is necessary, and it is sometimes surprisingly difficult to fix the vessel and compress it. It is to be remembered that in the limbs the course of the arteries is, in the main, nearly straight and lengthwise of the limb, so that all that it is necessary to do in case of severe arterial hæmorrhage is to feel along the outer edge of the wound, an inch or so from it, pressing the finger down deeply until the artery is felt pulsating, and then compress it against the nearest bone.

239. Permanent Arrest of Hæmorrhage.—When the hæmorrhage has thus been temporarily arrested, the pressure must in some way be kept up until there is no danger of a fresh burst on its removal. Nature's method of permanently arresting bleeding is this: It has been before stated that blood will coagulate not only outside the vessels, but inside them, if its free motion is interfered with. Now, if from the contraction of the vessels, or coagulation of the blood which has flowed from them, or in consequence of pressure artificially applied, or for any other reason, the current of blood in the artery is stopped at the severed end, the fibrin of the blood begins very soon to coagulate inside the vessel, and this coagulation extends from the cut end backward toward the heart to a point where the circulation becomes free and unobstructed, that is, to the point at which some branch artery is given off from the wounded one (Fig. 48). Thus the coagulum (or clot) formed will vary, according to circumstances, from an eighth of an inch or less to an inch in length.   If the compression or other obstruction continues long enough to allow of it, this coagulum becomes firmly · attached to the inside of the vessel, and forms a plug which effectually and permanently closes it, and, as the wound heals, this plug and the walls of the now useless artery become gradually absorbed until a mere thread remains, and even this may disappear so as to leave no trace of the vessel which formerly existed.   But it is to be considered that during the formation of this plug, and its attachment to the walls of the vessel, it has to re-

Fig. 48.—Clot in an artery that has been tied.

ceive the impulse of the blood continually driven against it seventy times a minute by the heart; and in large vessels this blow is a very strong one; so that, if the pressure be removed from an artery too soon, even after this coagulum is formed, the impulse of the current of blood may be sufficient to drive out the plug, and bleeding will begin again.   This danger is absent in venous hæmorrhage, and it is a general rule that if such hæmorrhage is once stopped it will not recur.   But in arteries this is a peril that must always be guarded against, and surgeons accomplish what they want by tying the end of the bleeding artery with a string.   Ligatures, made expressly for this purpose, are used, and although they are finally discharged from the wound with the secretions, showing that they have cut through the walls of the vessel around which they were tied, they remain long enough to accomplish the purpose.* A person's finger soon becomes exhausted by continual exertion in maintaining pressure, but a ligature keeps the artery closed for several days before it comes away, and thus ample time is afforded for the permanent closing of the vessel in the way above described. The application of a ligature re quires special knowledge and skill, and should only be attempted by a surgeon.

240. Recapitulation.—The means at everybody's hand, then, for arresting hæmorrhage are these:

1. The application of cold by water, ice, or air.

2. Pressure by the finger, thumb, bandage, or in any way that suggests itself.

(*a.*) If the wound be small, pressure on the wound itself.

(*b.*) If the wound be large, and the bleeding from the veins, pressure on the side farthest from the heart, or by plugging the wound full of some soft material. If the blood comes from an artery, pressure on the side nearest the heart.

(*c.*) Pressure to be kept up until the bleeding stops, or until some means can be applied to make the pressure permanent.

3. Styptics or astringents. These are to be used chiefly where pressure can not be applied, and after the application of cold has proved insufficient.

241. Wounds of the Extremities.—Wounds of the *extremities* often bleed profusely. Usually, direct pressure on the spot of the wound will suffice

---

* Surgeons are now in the habit of using ligatures made of some animal material, such as catgut or chamois-leather, which does not irritate the wounded parts, and is gradually absorbed, so that the wound can be completely closed when first dressed.

to check it. If, however, the wounded vessel is so large that it bleeds in spite of this, the main artery must be compressed above the wound. It has been already shown that the artery is a single trunk only in the upper part of each limb—i. e., from the shoulder to the elbow in the arm, and from the groin to the space behind the knee in the lower extremity. Somewhere in this course, then, the vessel must be compressed against the bone beneath; for below the elbow and knee the artery divides, and the branches are so sit-uated that they can not be compressed. The brachial and femoral arteries—that is, the arteries of the upper arm and thigh—are so large and strong, and receive such a strong impulse from the heart, besides being in a measure protected by the surrounding tissues, that they can not be compressed by the fingers to any advan-tage. The best method yet devised is the use of a *knotted handkerchief* or other bandage, a rope or cord be-ing so small as to cut the flesh, and therefore unsuit-able (Fig. 49). A handker-

Fig. 49.—Manner of compressing an artery with a handkerchief and stick.

chief should be tied loosely around the limb, with a hard knot over the artery. Immediately under-neath the knot should be placed another handker-chief, folded so as to form a pad about two inches wide and three long, to keep the knot from bruising the flesh when the handkerchief is tightened (Fig. 50). The handkerchief should be loosely tied in the

FIG. 50.—Direct compression of a wound by means of what surgeons call a graduated compress, made of pads of cloth, folded in different sizes, with the largest one on top.

first place, so as to allow a stick or rod to be introduced between it and the skin for the purpose of twisting it. The stick or lever should be placed at some distance from the compressing knot, and it is better to put it on the outside of the limb, where nothing will interfere with the twisting. The stick is now to be twisted round and round, making the handkerchief tighter and tighter and the knot press down deeper and deeper, until finally the artery will be compressed between the knot and the bone, and the hæmorrhage will cease. This method of compression is effectual and easy of application, but of course is only temporary, for such complete encircling of a limb and entire stoppage of the circulation produces gangrene if too long continued.

242. Fainting.—When, for any reason, the supply of blood to the brain is insufficient for its nutrition, the person *faints*. In our ordinary erect position, the blood has to be driven upward by the heart for a foot or more, in opposition to the force of gravity. In a fainting person there is not power enough to do this, and we must relieve the heart of a cer-

tain amount of its burden. We accomplish this end
by laying the person flat on his back, without raising the head by a pillow or rest of any kind. In
this position the blood readily reaches the brain,
and that organ rapidly recovers its functions.

243. Shortness of Breath.—Shortness of breath
always indicates that the blood contains too little
oxygen. When we are short of breath from exercise, it is due to the fact that our bodies have made
more waste material than usual, which the blood
has been unable to get rid of, and also that the oxygen taken in from the lungs is insufficient for the
needs of the wasting tissues. In other words, we
are wearing out. Then we begin to breathe faster,
and thus try to get rid of more waste and take in
more fresh material. But, if the exercise which produces the excess of waste be continued, the time
comes when we can not breathe rapidly enough to
dispose of it, the body becomes limp, and we are
forced to rest and recover. After we have completely ceased our exertions the rapidity of breathing continues for a time, making the supply much
greater than the waste, and gradually expelling the
latter from the body until the balance between the
two has become even again and the parts are all in
their natural condition. In diseases of the lungs
which render parts of these organs useless, such
as pneumonia, consumption, etc., the same effect is
produced. When we have only half as much lung
to breathe with, we have to breathe much faster to
make up the deficiency.

PART IV.

ORGANS OF CO-ORDINATION.

CHAPTER I.

NERVE-SUBSTANCE.

244. Difficulty of Investigation.—The *nervous system* is less thoroughly understood than almost any other portion of the body. The difficulties of investigation are enormous, and its functions are so intimately connected with the phenomena of conscious life that, with the exception of a few facts on which all observers are agreed, the truth is buried under a mass of conflicting theories, which the earnest and tireless efforts of patient workers all over the world are not yet able to remove. We shall occupy ourselves mainly with what are accepted facts.

245. The Two Divisions of the Nervous System.—We have seen, in previous chapters, that certain operations constantly go on in our bodies, not only without our willing them, but without our consciousness. Such are the processes of digestion, circulation, etc. Our voluntary acts, those which are the result of and accompanied by consciousness, constitute, in fact, the smallest part of what goes on within us. Now, the organs which are un-

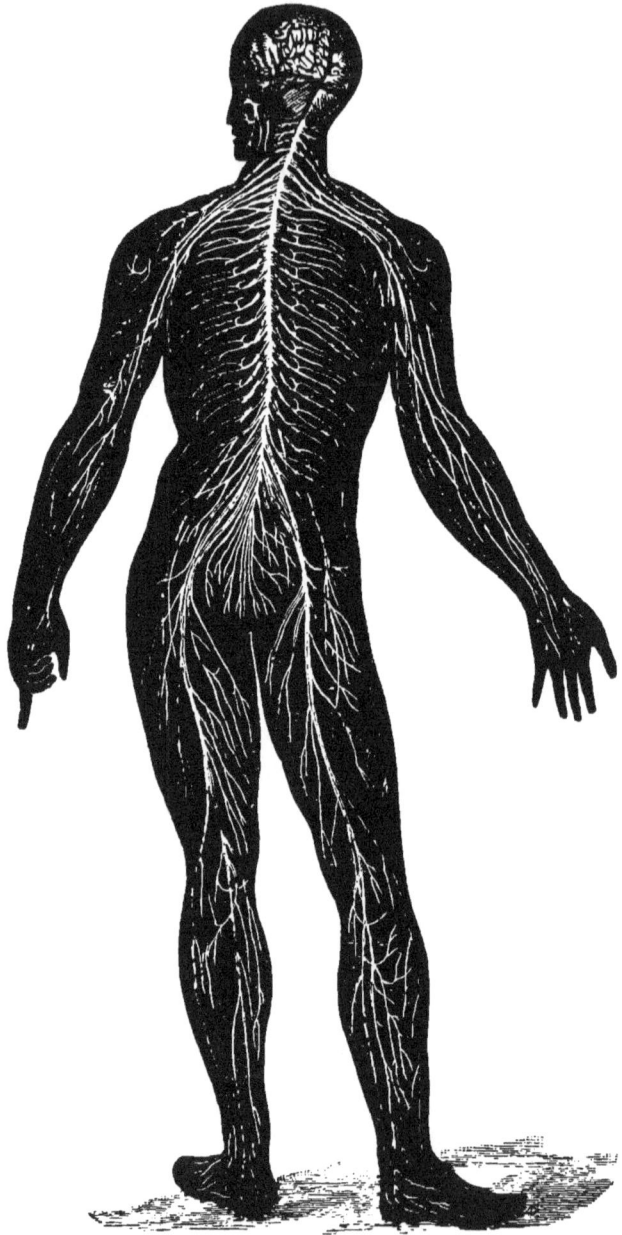

FIG. B.—The cerebro-spinal system of nerves.

der voluntary control are called the *organs of animal life*, while those which are beyond our control are known as those of *organic life*. Corresponding to these two divisions of the body are two great divisions of the nervous system—one of which has charge of the organs of animal life, and is called the *cerebro-spinal system*, while the other regulates the processes of organic life, and is called the *sympathetic system*. The name of the former indicates the fact that it comprises the *cer'ebrum*, or brain, and the *spinal cord*, with the nerves proceeding from them ; and the name of the latter suggests one of the principal characteristics of that portion of the nervous system, viz., that of inducing and regulating a sympathy between different organs, as will be hereafter explained.

246. Nervous Tissue.—Nervous tissue, wherever it is found, is made up of so-called *white* matter, or *gray* matter, or of both combined.

247. Nerve-Fibers.—The *white* matter is found, on close examination, to be made up of slender fibers, running parallel to each other (Fig. 51). They vary, according to their situation, from the $\frac{1}{15000}$ to the $\frac{1}{12000}$ of an inch in diameter, are nearly cylindrical in shape, and have for an outer layer a thin, delicate membrane,

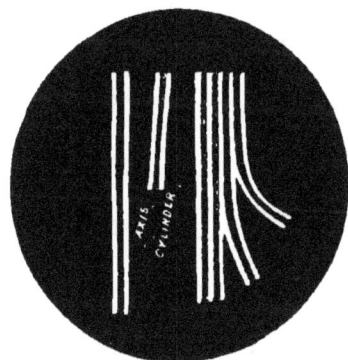

FIG. 51.—Nerve-fibers.

which serves to protect the nerve-substance and retain it in shape. Just inside the membranous covering is an almost trans-

parent material, which, after death, appears to coagulate and becomes whitish and slightly granular.
This is called the *my'elin*, or, by others, the *white
substance of Schwann*, from the man who first described it.   In the center of the whole runs a slender thread of transparent, very finely granular matter, called the *axis-cylinder*.   This latter substance,
in all probability, serves for the actual conduction
of the nervous influence, whatever it may be, and
the white substance of Schwann, or myelin, probably acts as an insulator.   These distinctions between the different portions of a nerve-fiber are not
visible in the living body, but are the result of
changes which take place after the nerve has been
separated from its connections, probably from a
sort of coagulation.

248. Nerve-Cells. — The *gray* matter does not
consist of fibers, but of
cells (Fig. 52) imbedded
in a mass of granular matter.   These *cells* vary in
size from $\frac{1}{200}$ to $\frac{1}{2000}$ of
an inch in diameter, and
each contains a *nucleus*
and *nucleolus*, usually very
distinctly marked.   Each
cell also has prolonga-
tions which extend from
its circumference in vari-

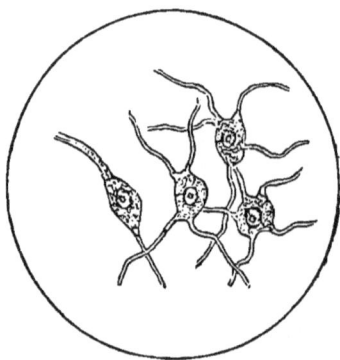

FIG. 52.—Nerve-cells.

ous directions, and are supposed in every instance
to connect either with a nerve-fiber or with some
other cell.   These prolongations vary in number
from one up to five or six, and, as they are traced
along their course with the aid of the microscope,

they are seen to divide and subdivide until they become too small to follow. It is supposed that every nerve-fiber is connected with such a cell.

The *nerve-fibers*, above described, constituting the white nerve-matter, are mere *conductors of the nerve-force*. They constitute the means of communication between the outside of the body and the nerve-centers. The *nerve-centers*, on the other hand, are made up of gray matter, and are the *originators of nerve-force*. So we see that the nervous system, like the rest of the body, is made up of cells and fibers, and that the essential part of the whole is the cell, the fiber playing a very subordinate *rôle*.

249. Structure of the Nerves.—A nerve which is large enough to be seen by the naked eye is composed of a great number of the fibers above described, lying side by side and bound together by a considerable amount of connective tissue. The connective tissue is greatest in amount in situations where the nerve is most exposed to injury—for example, in the limbs. In the brain, on the other hand, and in the spinal cord, where the nerve-substance is protected by a strong bony covering, the amount of connective tissue is scanty and the fibers are very small.

The white nerve-matter constitutes much the larger part of the whole nervous system. It forms the great nerve-trunks which go to the limbs and to the exterior of the body in every direction, and also forms the greater part of the brain and spinal cord.

250. Ganglia.—The gray matter, however, forms an important part of the brain and spinal cord, and also of various small nerve-centers, called *ganglia*, which are scattered throughout the body. In fact, every collection of gray matter, which is separated from other masses of gray matter by intervening white matter, is called a *ganglion*, and so even the different parts of the brain are included under this name.

251. Function of Nerve-Fibers.—The function of the nerve-fibers, as has been stated, is to convey impressions from one end to the other. Each nerve-fiber, even in the largest bundle, is completely isolated from all others, and runs an uninterrupted course from the cell, where it takes its origin, to the point of its termination. At its termination, it receives an impression, and this impression acts as a stimulus, which produces some molecular change through the whole length of the nerve, and eventually in the cell from which it comes. How this communication is accomplished is not yet known. Some change is produced in the nerve, which is not to be detected in any way excepting by its effect at the farther extremity.* In a similar manner, a change originating in a nerve-cell is transmitted along the fiber or fibers connected with it, and manifests itself at the termination of the fiber in some peculiar manner. Some nerve-fibers thus convey impressions from the outside inward to the nerve-centers, while others convey them from the nerve-centers outward to the exterior of the body. An intelligible illustration of this is to be seen in

* It has been found by the use of delicate instruments (galvanometers) that there is a constant electrical current in all nerves in a state of rest, supposed to be due to the nutritive changes taking place in the nerves. When the nerve is stimulated, however, so that the natural nerve-current is conveyed along the nerve, the electrical current is diminished in force.

the phenomena of ordinary sensation and muscular contraction. The prick of a pin stimulates the extremity of a nerve-fiber, or perhaps of several. The current originated by this stimulus travels along the course of the nerve to the nerve-center, and produces what we call *sensation*. Then the nerve-center, acted upon by this stimulus, sends out a certain amount of nerve-force along another nerve-fiber. This impulse follows the course of the nerve down to the muscular fiber in which it terminates, and there produces contraction of the muscle, and removal of the part which has been pricked, out of harm's way. This process, so simple to describe, so plain in its gross outline, is probably a very complex one, and it is not at all understood. When it takes place, as it often does, without any consciousness on the part of the person, it is called *reflex action*, because the result of the first stimulus, which produced the sensation, is reflected, as it were, along the second nerve to the muscle.

252. Rapidity of Nerve-Current.—It seems, at first thought, as if the reaction to a nervous stimulus were instantaneous. If the finger be hurt, it is jerked away so quickly that there seems to be no interval between the injury and the action. And yet it is evident that, if the received theory be true, viz., that a current must go from the finger to the brain in order to be felt, and from the brain back to the muscle in order to produce muscular contraction, there must be a lapse of time. Most people would probably say, however, that the nerve-force must travel as rapidly as electricity, and hence the time required for such a circuit would be inappreciably short. It has been shown, however, by ingenious and beautiful experiments, that the nerve-force does not travel over one hundred and ten feet (thirty-three metres) per second, and hence to go from the foot to the brain would require at least one twentieth of a second, and the same to return. Thus an injury to the foot would not be followed by voluntary muscular contraction until at least one tenth of a second had elapsed. Electricity, on the other hand, travels with almost inconceivable rapidity.

253. Exhaustion of Nerves.—The transmission of an impulse along a nerve-fiber not only requires time, but exhausts the nerve. We have previously shown that continued muscular contraction exhausts the muscular fiber, and that every muscle must have rest and nourishment, in order to maintain itself in health and vigor. It is the same with the nerves, but they recover, when exhausted, much more slowly than the muscles, and require a longer period of rest.

254. Nerve-Fibers as Conductors of Force.— The fact that nerve-fibers are mere conductors of nerve-force, and do not originate it, is shown by cutting them in two. All sensation and motion then cease in the portion of the body supplied by the nerve which has been cut. After some time, however, such injuries become healed, the cut extremities of the nerve unite, and the powers of sensation and motion return.

255. Gray Matter originates Force.—It has been already stated that the gray matter originates force. This is indicated by the fact that all the nerve-fibers end in collections of gray matter, and that in the natural condition no nerve-fiber conducts an impulse in either direction, inward or outward, unless it be directly connected with gray matter. What the force may be which resides in the nerve-cells, and what changes accompany their action, we do not know, and possibly never shall know. It has been often compared to electricity, and was once supposed to be identical with it, but it has been plainly demonstrated to be different. The diminution of the electrical current during the passage of the natural nerve-current shows this, and, if a nerve be tightly bound, the transmission of the nervous impulse is prevented, while the electrical current will pass through the constricted nerve without appearing to meet with resistance.

256. Difference between Cerebro-spinal and Sympathetic Nerves.—The cerebro-spinal nervous system and the sympathetic nervous system bear a close relation to the voluntary and involuntary muscular tissues respectively. The *voluntary* muscles are supplied by nerves from the *cerebro-spinal* system, and the *involuntary* largely by the *sympathetic*, and the same differences, that we have observed between the two kinds of muscular tissue in their manner of contraction, are also found in the two systems of nerves. A stimulus applied to any portion of the cerebro-spinal system produces an immediate response, while irritation of any portion of the sympathetic system only produces an effect after the lapse of an appreciable time.

The only portion of the nervous system over which we have any control whatever, and which stands in any relation to our consciousness, is the cerebro-spinal system (Fig. 53). This system, constituting by far the greater bulk of our nervous ap-

14

paratus, comprises the brain and the spinal cord,
together with all the nerves which take their origin

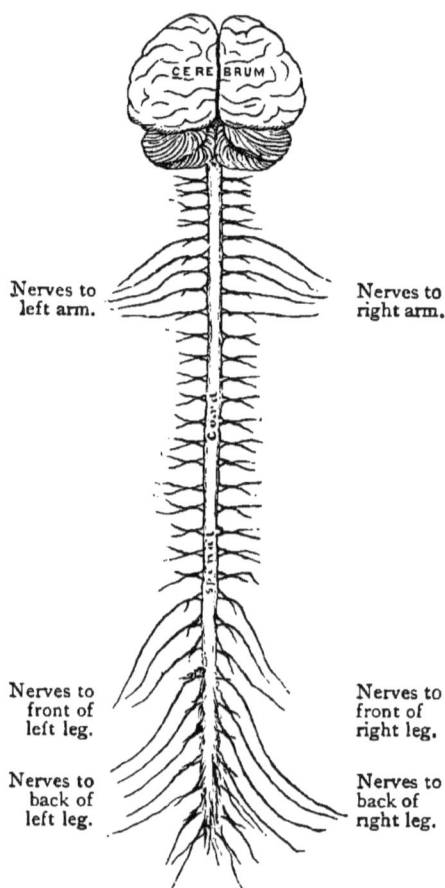

Nerves to left arm.   Nerves to right arm.

Nerves to front of left leg.   Nerves to front of right leg.

Nerves to back of left leg.   Nerves to back of right leg.

FIG. 53.—Brain and spinal cord, with the thirty-one pairs of spinal nerves.

in these organs. The brain is a very complex organ, being composed of several different ganglia,
each of which has its own peculiar functions. In
describing the nervous system it will be convenient,
therefore, to begin with the sympathetic system, as
being the simplest in its structure and functions.

CHAPTER II.

257. Structure of the Sympathetic System.—The *sympathetic system* consists of a double chain of nervous ganglia in the head, neck, and trunk, sending fibres to various organs and to the blood-vessels throughout the body (Fig. 54). The arms and legs

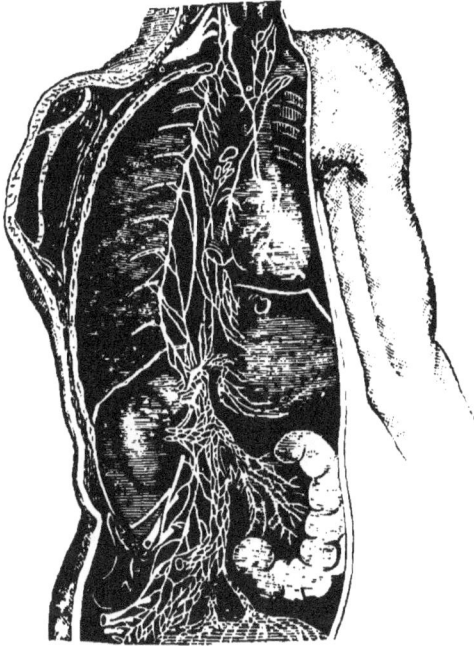

FIG. 54.—The sympathetic system of nerves in the trunk.

are organs of animal life, and are supplied with cere-bro-spinal nerves. The ganglia of the sympathetic vary very much in size, some being only visible with the microscope, and others as large as a pea, or, rarely, even larger. They are composed, as has been said, of gray nerve-matter, and are connected with each other and with the cerebro-spinal nerves by means of communicating fibers. Some of these fibers are of the ordinary white matter, while others are transparent and grayish in color, and appear to consist of an axis-cylinder alone, without any surrounding myelin. The sympathetic ganglia all lie very deep in the cavities of the body, in the vicinity of and surrounding the important organs,* whose functions they control, and it is very difficult to get at them for purposes of experiment. The consequence of this is, that very little has been learned about the real action of the sympathetic system, and many of its functions can at present only be conjectured.

258. Sluggish Action of Sympathetic Nerves.—The sympathetic nerves have been proved to be capable of conveying both sensory and motor impulses; but these properties are very slow in manifesting themselves, in marked contrast to the behavior of the cerebro-spinal nerves. If the extremity of a sympathetic nerve be irritated, it is only after a considerable time that the nervous center is af-

---

* The large sympathetic ganglia, called the solar plexus, or the semi-lunar ganglia (from their shape), lying behind the pit of the stomach, constitute what has been sometimes called the abdominal brain. A blow in this region is very dangerous, for, if it is powerful enough to paralyze these ganglia by the shock, it is more certain to cause instant death than the severest blow on the head.

tected, and the reflex motion is very slow in its
appearance. This fact is illustrated in the inflam.
mations and congestions of internal organs. The
sympathetic system mainly furnishes the nervous
supply of the organs of digestion and secretion. If
any irritation, like the cold air of a draught for instance, affects the body, the result, such as internal
congestions, etc., only appear after some time has
elapsed. An exposure to injurious influences may
produce a serious disease through the sympathetic
system, and yet the disease may not declare itself
for ten or fifteen or even twenty-four hours after
the exposure.

259. Contraction of the Pupil.—Various familiar
phenomena show this sluggish action of the sympathetic nerves. One of the most obvious is the contraction of the pupil of the eye under exposure to
light. When a strong light is thrown into the eye
the pupil grows smaller and shuts out the excess of
light. On the other hand, when the supply of light
is diminished, the pupil enlarges so as to admit
more. Now, these changes in the size of the pupil
take place very slowly. When we go from a light
room into a dark one it is several seconds, sometimes a minute, before we can see anything, simply
because the pupil does not admit enough light and
requires time to enlarge sufficiently. When we go
from a dark room to a very light one we are dazzled; the pupil is too large and admits too much
light, and we are obliged to shade our eyes until
contraction has taken place. These phenomena are
the result of impressions made on the sympathetic
system, and serve as an excellent illustration of the
slowness and precision of its action, as well as of the

fact that its functions are entirely beyond our conscious control.*

260. Effect of dividing a Sympathetic Nerve.— One of the most remarkable facts relating to the sympathetic system was discovered by the celebrated physiologist, Bernard, and has been substantiated since by many careful observations on persons in whom different sympathetic nerves have been incapacitated by disease or severed during a surgical operation. If, for instance, the sympathetic nerve running to the ear be divided, the ear becomes red and hot, and the blood-vessels are seen to be enlarged and much fuller of blood than they usually are. There is no stagnation of the circulation, and, after the lapse of a sufficient time, three or four weeks, if the divided ends of the nerve unite, the ear returns to its natural condition. If, after the nerve has been divided, as above, while the ear is in its red and hot condition, the end of the nerve nearest the ear be irritated by a galvanic current, the blood-vessels contract, the unusual amount of blood is expelled from them, and the ear resumes its ordinary temperature and color. If the cause of irritation be removed, the redness and heat return. These facts seemed to indicate that the supply of blood in a part was regulated by the sympathetic system of nerves, a supposition which has been confirmed many times over.

---

* The iris—i. e., the muscular curtain surrounding the pupil—is supplied with nerves from the cerebro-spinal system as well as from the sympathetic. The relative functions of the two sets of fibers are not well understood, but the sluggish and involuntary character of the changes in the size of the pupil shows that the influence of the sympathetic nerves predominates.

**261. Influence of the Sympathetic Nerves on Se-
cretion.**—It has been found that, if the sympathetic
nerve supplying one of the salivary glands be divided, an increased supply of blood takes place and
increased secretion by the gland follows. The submaxillary gland being treated thus, an increased
flow of saliva takes place. A similar effect is produced by galvanic irritation of the nerve of taste.
Any savory substance in the mouth, then, irritating the nerve of taste, gives rise to a reflex action
through the nerves supplying the salivary glands,
and produces a flow of saliva. This is what we
call "feeling the mouth water." The same effect may even be produced by purely emotional
causes, as the mere sight of an appetizing article
of food.

**262. Effect of Emotion on the Sympathetic
Nerves.**—Blushing is another phenomenon depending on the control of the sympathetic nerves over
the blood-vessels. The emotion of shame produces
a temporary paralysis of the sympathetic nerves,
and gives rise to the same effects as a division of the
nerve. Blood rushes to the superficial blood-vessels
and they become redder and hotter than they ordinarily are. The blush is more evident in the cheeks
than elsewhere, because the skin is thinner there,
and the blood-vessels more numerous ; but the
blush extends, in reality, much more widely than is
commonly supposed, and covers a large portion of
the surface of the body. The peculiar characteristics of the sympathetic system are to be plainly
discerned in the blush. It is not instantaneous, but
comes on slowly after an indignity, gradually rises
to its greatest height, and then gradually disap-

pears. It is also beyond the control of the will. If a person really feels the emotion, he can not restrain the blush. The only means of preventing it lies in such a constant schooling of the mind that feelings of shame, modesty, insulted dignity, etc., shall not be felt. When a person has arrived at this point of self-control he will not blush. But who would wish to purchase exemption at such a price? Many persons are ashamed of blushing, but who would be ashamed of being ashamed? The two can not be separated, and one who has lost the sense of shame entirely has lost much of what commends us to the sympathy and respect of our fellows.

263. The Vaso-motor Nerves. — The nerves which thus control the supply of blood to the blood-vessels are called the *vaso-motor* nerves. They have been already referred to in another part of this book. They have received the name of vaso-motor because they control the motion of the walls of the vessels (*vasa*) producing contraction or relaxation. Although they are chiefly of the sympathetic system, they receive fibers from the cerebro-spinal system. The two systems are not entirely distinct from each other, but there are certain functions in which they are so widely different that we can say with certainty this action belongs to the cerebro-spinal system, or that one to the sympathetic; while between these extremes are all sorts of gradations, so that in many cases we are unable to distinguish the characteristics of either system, and must say of them that we do not know how they are produced or that both systems probably unite their functions.

264. Influence of the Sympathetic System on Di-

gestion.—The process of digestion is presided over and regulated mainly by the sympathetic nervous system. The introduction of food into the stomach stimulates the nervous ganglia in the abdomen, and through their influence is followed by all the phenomena of secretion, muscular movements, and absorption, which have been shown to accompany and form a part of the process of digestion. All of these phenomena are beyond our control, but, as has been shown in the case of the blush, the nervous supply of the sympathetic is favorably or unfavorably affected by strong emotions. And it is through these ganglia and their connections that anger, fear, or other depressing emotions, produce such an injurious effect on the digestive organs during and after a meal.

265. Effect of Cold on the Sympathetic System.—It is through the reflex action of these nerves, also, that hæmorrhage can often be checked by the application of cold, and that inflammation of internal organs may result from the same cause. The mucous membranes, for the same reason, seem to be more exposed to such inflammations than any other portion of the organism. Hence, exposure to a cold draught is very apt to bring on such a disorder. In one person it may affect one mucous membrane, in another another, according to the relative condition of the membranes at the moment. After such an exposure, one person may have a severe cold in the head, another a bronchitis, another a sore throat, another a diarrhœa or inflammation of the mucous membrane of the intestines.* All of these results

---

* Many persons, instead of catching cold in the nose or throat after exposure, have their bowels affected, and the resulting catarrh causes a

are beyond our control, and the only way to pre-vent them is to keep out of danger. When a person is in health, food introduced into the stomach will inevitably be digested, whether he wills it or not, and, if he undergoes certain exposures, he will take cold, in like manner, whether he wills it or not.*

266. Exhaustion of the Sympathetic System.— As the sympathetic system is so intimately con-nected with the processes of nutrition, and indeed with all the functions of those organs which main-tain us in life, without our consciousness, it is evi-dent that it should be well taken care of and nothing done to cripple it. We have, then, to bear in mind that nerves require rest as well as any other part of the body, that they easily become exhausted, and that they require a longer period for recuperation than the muscles do. The surest sign, perhaps, of exhaustion of the sympathetic system will be found

diarrhœa. When a person is specially liable to this affection, he should wear a thick flannel bandage around the abdomen, which will furnish almost entire relief from such attacks.

* A cold is generally the result of the sudden chilling of some part or the whole of the surface of the body. The first effect of such a chill is to contract the blood-vessels on the surface and overfill those that lie deeper, and thus to cause congestions of the internal organs. Such an ef-fect seems to be especially apt to follow a cold draught striking the back of the head or neck or the ankles. Now, active muscular exercise tends to drive the blood to the surface, and is therefore a natural preventive of colds. It tends to keep the blood at the surface, where a conges-tion will be relieved by perspiration, and prevents its accumulation in the deeper organs, where it is more likely to cause trouble. If expos-ure to cold, therefore, can not be avoided, we must keep up a brisk circulation by physical exercise. Wet clothing should be changed for dry as soon as possible, for so long as it remains in contact with the body it absorbs a great amount of heat from it and keeps up a constant chilling of the surface. When a person is not engaged in physical ex-ertion, he should avoid draughts.

in *failure of nutrition.* If a person grows thin and pale and languid, notwithstanding a good supply of food, or if his food be not appropriated by the body—in short, if his nutrition become impaired, and no organic disease be discoverable—it is probably the result of nervous exhaustion, and means should be taken to relieve the overtasked organs. Rest will accomplish wonders in many cases of illness, especially where there is no disease of any particular organ to be detected.

267. Structure of the Spinal Cord.—The *spinal cord* is that portion of the nervous system which lies within the spinal canal. It extends from the junction of the spinal column with the skull, down to the loins, and is about a foot and a half long and a little less than half an inch in diameter. It weighs about an ounce and a half. It does not occupy the whole spinal canal, but the space around it is filled with membrane, blood-vessels, and nerves. At its upper extremity it passes into the brain, and at its lower, just below the last rib, it divides into a bundle of small nerves, presenting very much the appearance of a tassel at the end of a thick cord.*

The spinal cord consists partly of white and partly of gray matter. The gray portion occupies the central portion of the cord, and is arranged somewhat in the form of the letter H, with the two upright marks curving outward at both ends (Fig. 55). The remainder of the cord, in which this gray substance is imbedded, is composed of white matter.

The spinal cord, throughout its whole length, is divided by fissures, one of which extends from be-

---

* This part of the spinal cord is called the *caud'a equi'na*—i. e., the *horse's tail*, which it somewhat resembles.

fore backward, and the other from behind forward, the two nearly meeting. They are, however, separated, and the halves of the cord kept in communication by a bridge of gray matter, which passes over from one to the other, forming the cross-line of the letter H.

268. The Spinal Nerves.—The spinal cord sends out nerves through its whole course. With the exception of the head and face, all the muscles and the skin of the body receive their nervous supply from this source. These nerves, thirty-one pairs in all, issue from the spinal column between the vertebræ through small openings.

269. Properties of the Spinal Nerves.—All parts of the body, which are supplied with spinal nerves, are endowed with two remarkable properties, *sensation* and *motion*. As long as the nervous supply remains in a natural and healthy condition, any portion of the body feels, and any portion can be moved from one place to another. These movements are sometimes voluntary and sometimes we are entirely unconscious of them, and it is only within a few years that physiology has been able to offer any explanation whatever of these facts.

270. Sensation and Motion.—When we consider attentively the property of sensation, we find that there are two great divisions of it, easily distinguishable and easily demonstrated, viz., *sensibility to pain* and *ordinary sensation*. When a person is brought under the influence of an anæsthetic, the sensibility to pain disappears before ordinary sensation is lost, and even while the individual is still conscious. In slight surgical operations this fact is often strikingly manifest. A tooth may be pulled, and the patient be conscious of every step of the proceeding, without feeling the slightest pain. Narcotics, too, may relieve pain, while the person who has taken them remains perfectly in possession of his senses. Physiologists have made other discriminations of sensation, which it is not necessary to mention here.

The phenomena of motion are familiar, and require no separate consideration.

271. Effect of dividing a Nerve.—If one of the spinal nerves be disabled, sensation and motion are abolished in the portion of the body supplied by that nerve. This shows that sensation and motion are both conveyed in the same nerve. But here a difficulty arises. Sensation implies the passage of a nerve-current from without inward, toward a nerve-center; whereas motion implies the passage of a current from within outward, from the nerve-center to the muscle. Can both these currents be transmitted at the same time, or can the axis-cylinder of the nerve-fiber be used for the passage of the nerve-current in either direction according to circumstances? This is the problem which waited until this century to be solved. The simplest explanation has been found to be the correct one. As every nerve is made up of many fibers, it would be natural to suppose that some of these fibers might serve as conductors of sensation, and others of motion. This was made all the more probable by the fact that, in many forms of disease, it was seen that the power of sensation might be abolished, while that of motion remained; or, on the other hand, the power of motion might be paralyzed, while that of sensation remained unimpaired.

And, in fact, the above explanation of these phenomena was found to be correct. The first who
demonstrated this appears to have been Sir Charles
Bell,* although Magendie † has strong counter-
claims to the honor. The observations by which
the truth was established were decisive, and have
been often repeated by other physiologists. They
are the following :

272. The Roots of the Spinal Nerves.—The spinal nerves do not take their origin from the spinal
cord as single trunks, but each one arises by two
roots. One of these roots is formed of fibers coming from the anterior portion of the cord, and is
called the *anterior root ;* the other is formed by
fibers from the posterior portion of the cord, and
is called the *posterior root ;* the latter, at a short distance from its source, passes through a small ganglion of gray matter. These two roots, anterior
and posterior, approach each other and unite in a
single cord just before leaving the spinal canal.
This cord, therefore, contains fibers from both portions of the spinal cord, and, after running a short
distance, it again divides into two nervous trunks,
one of which supplies the front of the body and the
other the back.

273. Division of the Anterior Root.—When the
anterior root of a spinal nerve is disabled (Fig. 55),
the portion of the body supplied by the nerve of
which this particular root forms a part, loses its

* Sir Charles Bell, a distinguished Scotch anatomist, physiologist,
and surgeon (1774–1842), Professor of Surgery in the University of
Edinburgh.

† François Magendie, a celebrated French physiologist (1783–1855),
Professor of Anatomy in the College of France.

power of motion, but preserves its sensibility unimpaired. If the portion of the body supplied by this

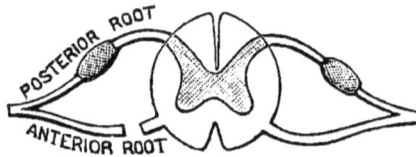

FIG. 55.—Cross section of spinal cord and roots of spinal nerves, with anterior root cut.

nerve be pinched or pricked, there are evident signs of pain, and the irritation is clearly perceived, but there is entire inability to move the muscles of the part. If, now, the ends of the nerve be irritated, the following results are obtained : On irritation of the end nearest the part supplied by the nerve, the muscles of that part become strongly contracted, or convulsed, while there is no indication of any sense of feeling whatever. If the end nearest the spinal cord be irritated, no effect whatever is produced. There are no convulsive movements of the muscles, and no indication of sensation. This seems to show that, in the fibers of the anterior root, the nervous impulse is transmitted from within outward, and gives rise to the phenomena of motion, while they have nothing to do with sensation.

274. Division of the Posterior Root.—If, now, instead of the anterior root, the *posterior root* is the injured one (Fig. 56), we find that the part supplied by the nerve has lost sensation, but retains the power of motion. If it be pinched, pricked, or injured in any way, no effect is produced, and no notice is taken of it whatever. This shows that

sensation in the part is abolished. If, now, the ends
of the root be irritated, we find the following results: On irritation of the end nearest the part, no
effect whatever is produced. If, on the other hand,
the end nearest the cord be irritated, there are immediate indications of pain, and the muscles sup-

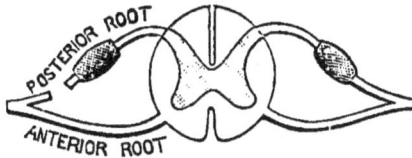

FIG. 56.—Cross section of spinal cord and roots of spinal nerves, with posterior root cut.

plied by the injured nerve are still able to contract
as powerfully as ever. This shows that the power of
motion has not been interfered with by the injury,
and it also shows that, in the posterior roots of the
spinal nerves, the nervous impulse is transmitted
from without inward, and has to do with sensation.

275. Two Kinds of Fibers in Each Nerve.—The
anterior roots, then, are composed of *motor* fibers,
and the *posterior* roots of *sensory* fibers. The two
unite, as has been shown, to form a single cord,
which afterward divides again to supply the front
and back of the body. Every spinal nerve in this
way is made up of both sensory and motor fibers,
or fibers which convey sensation and fibers which
convey motion; and thus, from the constitution of
the nerve, it is easy to see how the powers of sensation or motion can be abolished separately, one
remaining intact while the other disappears.

276. Relation of the Spinal Cord to the Brain.—

15

These properties of sensation and motion have been considered only in their relations to consciousness. The sensations have been spoken of as felt, and the movements as voluntary. In order for this to be so, there must be some connection between the nerves and the brain, for consciousness has its seat in the latter, as we shall hereafter see. And in fact it has been proved that the nerve-fibers, after entering the spinal cord, pass upward in its interior toward the brain. It is probable that the white matter of the cord is largely composed of fibers which connect the brain in this way with the exterior of the body.

As the nerve-fibers approach the brain, however, at the very summit or upper extremity of the spinal cord, they cross over from one side of the cord to the other, the fibers from the right side passing over to the left, and those from the left going over to the right. The crossing of the motor fibers only takes place at this point, while the sensitive fibers cross to the opposite side of the cord soon after joining it. The consequence of these facts is, that the right side of the body comes into direct communication with the left side of the brain, and the left side with the right brain. An injury to the brain, therefore, on either side, produces paralysis of the opposite side of the body. An apoplexy* affecting the *right* side of the brain brings on paralysis of the *left* side of the body ; and conversely, if we see a person whose *left* side is paralyzed, we

---

* Apoplexy, from the Greek, meaning a *sudden stroke*. It is caused by the giving way of some blood-vessel in the brain, and the blood escaping into, compressing, and tearing apart the delicate nerve-tissue, causes paralysis and often entire unconsciousness, or death.

look for the injury which has caused it on the *right* side of the head.

277. Sensations referred to Extremity of Nerve. —It has already been said that the nerve-fibers act merely as conductors of an impulse, and do not themselves originate nervous force. In consequence of this, an injury to any portion of a nerve always produces the same effect as if it were at its extremity. The pain which results from any impression on a nerve-fiber is always referred by the nerve-center which receives it to the termination of that particular nerve. And this fact clearly explains many things that once were not understood. The ulnar nerve furnishes a good illustration of this. At the inside of the back of the elbow are two bony projections, with a hollow canal between them. At the bottom of this canal, not far below the skin, runs a large nerve called the *ulnar nerve*, because it supplies the ulnar side of the fore-arm and hand. If this nerve be pressed or injured in any way, a tingling sensation is felt in the little finger and the adjoining side of the ring-finger, the parts supplied by this nerve.* Striking examples of the same phenomenon are often seen in persons who have lost limbs by amputation. If the end of the nerve which remains in the stump be irritated in any way, as by the contraction of the scar, the person feels the irritation precisely as if it were in the lost limb. If any person's arm were cut off, for instance, just below

---

* The ulnar nerve supplies the little finger, the adjoining side of the ring-finger, and that side of the hand as far as the wrist. Pressure upon it, therefore, interferes with the nervous supply of those parts, and they tingle and become numb. The queerness of the sensation produced has caused this nerve to be commonly known as the crazy-bone, or the funny-bone, although it is not a bone at all.

the elbow, and the stump of the ulnar nerve were irritated, he would feel the tingling sensation in the little and ring fingers of his hand just as much as if the hand were still attached to the arm. The sensation is precisely the same, and can only be corrected by the sense of sight or by the individual's own memory or reason. It is this fact which has given rise to the numerous accounts of persons who have felt people maltreating their amputated limbs, when they perhaps were already decayed, or were hundreds of miles away.

278. Reflex Action of the Spinal Cord.—All the phenomena which have been described might occur just as well if the spinal cord were merely a large bundle of white fibers running down from the brain, and sending out branches to different parts of the body. It has, however, a considerable amount of gray nerve-matter in its interior, and, as we have seen that this form of nerve-matter originates force, we are prepared to infer that the spinal cord can originate movements and receive sensations without any connection with or dependence on the brain. And this is found actually to be the case. If the head of a frog be cut entirely off, leaving the body otherwise uninjured, sensation and motion are still manifested in the trunk and limbs. If the toe be pinched, the leg will be drawn up out of the way. This is the same effect that is produced when the animal is entire, with this exception: there is no attempt to escape from further injury. In other words, there is no brain-action, no reasoning. The irritation sends a nerve-current along the sensory nerve to the spinal cord. There it produces certain changes in the cells of the gray matter, and from them another current is sent outward along the motor nerve to the muscles, and movement results. If the nerve be severed, no such effect can be produced. If the spinal cord be destroyed, the nerves remaining intact, no such effect will occur. These facts show that the cord itself has the property of receiving sensations and sending out motor impulses. This action of the spinal cord, by which it produces a movement in response to and in consequence of a sensation, is called *reflex action.*

279. Diseases of the Spinal Cord.—Experiments like the above, of course, can not be performed on human beings, but the course of nature often performs them for us. Diseases are not uncommon in which the spinal cord is completely cut across in some portion of its length, so that there is no nervous communication between the brain and that part of the body below the seat of disease or injury. In such cases we have paralysis of the kind called by physicians *paraplegia*—i. e., the lower portion of the body is no longer under control of the brain, and no longer capable of voluntary movements. It is also entirely beyond the domain of consciousness; injuries to the paralyzed part cause no sensation in the brain. The separation of such a part of the body from all relation to consciousness, whether of sensation or of motion, is as complete as if it formed part of another body. Notwithstanding this fact, we often find the spinal cord below the seat of injury to be in the same condition as that of the decapitated frog. If the disease or injury has been confined to a particular part of the cord, and the portion below, with its nerves, be still in a healthy condition, excepting that it is separated from the brain, a stimulus applied to any of the nerves will produce a response which is immediate and decided. A person in this condition looks on and sees his feet pinched, or pricked, or burned, and sees the irritation followed by violent convulsive kicks and thrusts of the limbs, without the slightest knowledge of what is going on, excepting what he receives through the sense of sight.

280. Automatic Actions.—Certain actions, which are at first voluntary, and only accomplished by great effort, become, after a time, so natural and easy, that they seem to be carried on unconsciously, through the action of the spinal cord. Thus, in the act of walking, a harmonious action of several muscles is required to keep the balance, an action which is at first learned only after prolonged and incessant effort. In after-life it becomes so easy as to be carried out without consciousness, as has been often shown in the case of soldiers, who have continued to walk in the ranks and keep up with their comrades on the march while fast asleep.

CHAPTER IV.

281. The Brain.—The *brain* and the *cerebrum* are often spoken of as if they were convertible terms. But this is not correct. The *brain* includes all that part of the nervous system which lies within the cavity of the skull. This great mass of nervous matter is made up of several distinct ganglia, which, to be

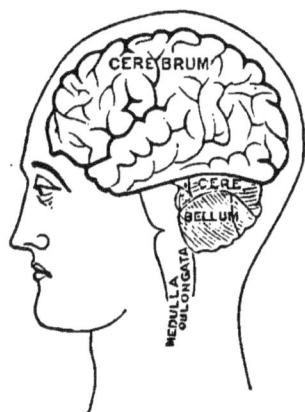

FIG. 57.—Diagram showing the position of the nervous centers in the head.

sure, are connected with one another and interdependent, and yet each of them has its own particular function which the others have not (Fig. 57). The *cerebrum* is merely one of these ganglia, and, although the highest in the scale as regards the character of its functions, it being the ganglion which principally gives man his pre-eminence over the lower animals, it is one of the least important as regards the mere preservation of life. In many of the lower animals it can be entirely removed, and the animal will live for months afterward. In the human being it is so plentifully supplied with blood-vessels that, if its removal should be attempted, the person would die of hæmorrhage. This has been the invariable result of experiments on the higher kinds of animals, as the dog or horse. But there is no doubt that, if the cerebrum could be removed without interfering with the circulation of the blood, even a human being would continue to live without it.

282. Structure of the Cerebrum.—The *cerebrum* (Fig. 58) is the largest part of the brain, forming as

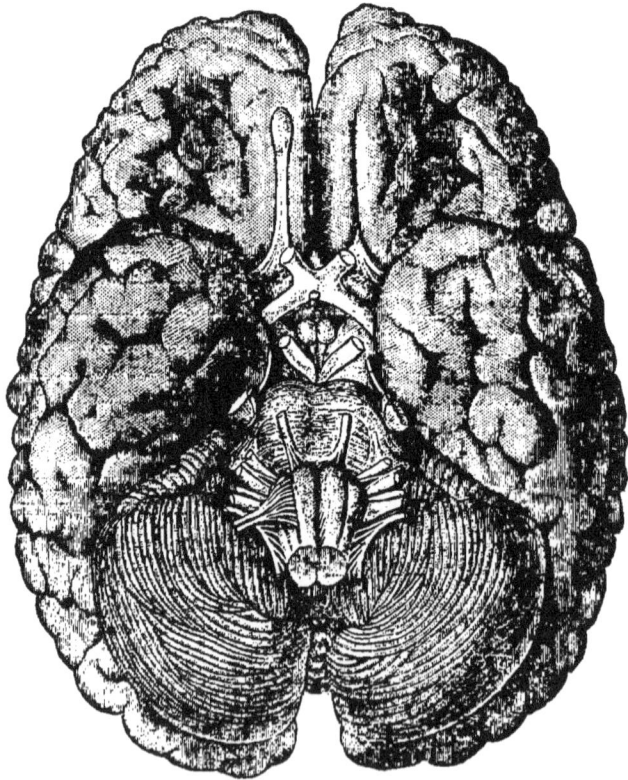

FIG. 58.—Under surface of the brain, showing the great complexity of its structure. At the lower part of the cut is the cerebellum.

it does about nine tenths of the whole mass within the skull. It consists of two halves, called *hemi-spheres*, the dividing fissure running from the front of the head to the back; but the halves are not entirely separated. On their under side, a bridge or mass of white fibers runs from one to the other, and serves as a means of nervous communication between them. The cerebrum is composed of both white and gray matter, the latter being spread over the surface, and the former filling up the interior, and forming the greater portion of the mass (Fig. 59).

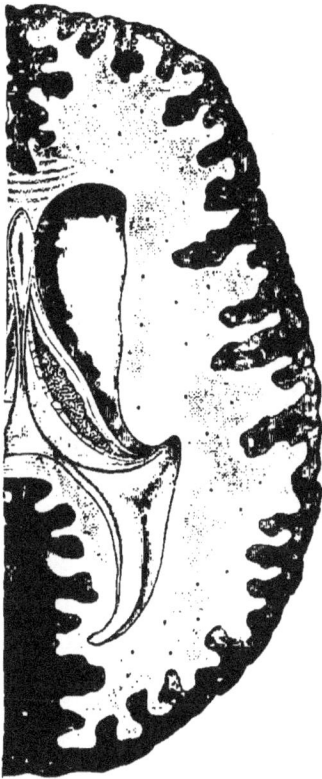

FIG. 59.—Section of the brain, showing the arrangement of the gray and white substance.

The arrangement of the nerve-matter of the cerebrum is very peculiar. It is formed into folds and wrinkles, somewhat like the meat of an English walnut. These folds are called the *convolutions* of the brain, and, as the gray matter follows closely the dipping of the surface between the convolutions, it is evident that the amount of gray matter is much greater than it would be if the surface of the brain were smooth.

It has been found by experiment that the nerve-matter composing the cerebrum is insensible to ordinary irritation. Either the gray or the white matter may be cut, pricked, burned with fire or caustic acids, without producing any sensation. What, then, is the function of the cerebrum?

283. Function of the Cerebrum.—It is now acknowledged by all who have paid any attention to the subject, that the *cerebrum* is the *organ of thought.* It is in this part of the brain that all ideas have their origin, and that all processes of reasoning take place. The evidence of this is indirect, but very conclusive, and may be briefly stated as follows:

284. Intelligence increases with Increase in Size of the Cerebrum.—The cerebrum increases in size in animals as we pass from the lower forms to the higher. The superiority of one animal over another in the scale of life depends mainly upon what we call intelligence. Now, we find that the intelligence of animals appears to increase in the same ratio that the size of the cerebrum increases in proportion to the rest of the body. In the fish, the cerebrum is very small, and his intelligence is very low. In the reptiles, the cerebrum is somewhat larger, and their intelligence as a class greater. In birds, the relative size of the cerebrum is considerably increased, and their intelligence is proportionally and noticeably higher. From birds we pass to quadrupeds, and here is a great increase in the size of the cerebrum, and a correspondingly finer intelligence. As we go up in the scale of quadrupeds, we find the size of the cerebrum increasing faster than the size of the body does, until, in the elephant, we find a brain weighing eight or ten pounds. The elephant and the whale are the only animals whose brains are larger than man's, and even their immense brains are much smaller relatively to the size of their bodies. The human brain varies in weight in different persons, but its average weight is fifty ounces, or about three pounds. These facts, then, seem to show a relationship between the cerebrum and the quality of intelligence, for the difference in weight of different brains is mainly due to the varying size of the cerebrum.

285. Size of Human Brains.—When we examine different brains among human beings, we find the same general law holding good. Other things being equal, a larger brain signifies a greater mind. Thus the brain of Cuvier,* the celebrated naturalist, weighed 64⅓ ounces; that of Abercrombie,† physician and philosopher, 63 ounces, and that of Dupuytren,‡ surgeon, 62½ ounces; while the brain of an idiot seldom exceeds thirty ounces in weight. Occasionally men have possessed great intelligence, and have been found after death to have very small brains. An example of this is the brain of a celebrated mineralogist (Haussmann), who is said to have been above medium stature. Its weight was only 43.24 ounces. In such cases, it is usually found that the convolutions are very deep and very much curved and turned about, so that there is a greater amount of gray matter packed away than would have been expected in so small a skull.

* Baron George Chrétien Léopold Frédéric Dagobert Cuvier, one of the most eminent of modern scientific men (1769–1832), Professor of Natural History in the College of France ; generally considered the founder of the science of comparative anatomy.

† John Abercrombie, a distinguished Scottish physician (1781–1844), especially celebrated for his metaphysical writings.

‡ Baron Guillaume Dupuytren (1777–1835), during his lifetime considered the most eminent surgeon in France ; Professor of Surgery at Paris.

286. Effect of Injuries of the Brain.—It has also been noticed, from the earliest times, that injuries of the brain produce unconsciousness and various other phenomena, which indicate that the organ of the mind has been injured; so that, by almost all men, the fact that the brain is the seat of the mind has been readily acknowledged. The determination of the particular portion of the brain which is the organ of mental processes has only been arrived at by long-continued study and observation.

287. Disease of the Human Cerebrum.—And here, as in the case of the spinal cord, we find experiments on human beings provided for us by Nature. When the cerebrum of a human being is diseased, we find that the mind is affected. This is noticeably the case in the disease commonly known as *softening of the brain.* The first symptoms in this distressing malady are mental ones. The person begins to lose his memory; his friends probably perceive it before he does himself; and from this first indication of intellectual feebleness the disease goes on, often without any disturbance of the nutrition of the body, until the intelligence is completely abolished, and the person's life is a blank. When the brain of such a person is examined after death, the cerebrum is the part found diseased, and we see that the gradual impairment and extinction of the mind, which we have been watching, is the result of physical disease.

288. Recapitulation.—To recapitulate; *the evidence that the cerebrum is the seat of the intelligence* is the following: *First,* we see that in the lower animals, as well as in man, the development of the

mind is proportionate to the development of the
cerebrum in comparison with the rest of the body.
Second, that if the cerebrum be destroyed, the mental faculties are the only ones lost. *Third*, when
the mind of a human being is impaired or lost,
we find a corresponding disease or injury of the
cerebrum.*

289. Structure of the Cerebellum.—The ganglion next in size to the cerebrum, and one whose
functions are of great importance, is the *cerebellum*,
or little brain (Figs. 57, 58). It is situated beneath
the back part of the cerebrum, occupying that portion of the skull immediately behind the ears. Its
average weight is five and a quarter ounces, about
one ninth of the whole weight of the brain. It is
composed of gray and white matter, arranged in
much the same way as in the cerebrum, excepting that the convolutions are smaller and dip in
deeper, so that in the cerebellum the proportion
of gray matter is much greater than in the cerebrum. This fact, of itself, would indicate its importance.

290. Function of the Cerebellum.—The exact
function of this part of the brain has not yet been

---

* Also, when the cerebrum is seriously injured, the mental faculties
are generally impaired. There have been some cases of recovery after
very severe injuries of the brain. In 1850 a young man in Vermont, a
farm-hand, was blasting rock, when a tamping-iron, three and a half
feet long and an inch and a quarter in diameter, was driven through
his head, entering the left cheek and coming out at the top of the head,
causing a great loss of brain-substance. He recovered completely from
the wound, but from a quiet, home-loving, honest lad, he developed
into a tricky, thieving fellow, of a roving disposition, lived in various
places in South America and the United States, was subject to epileptic fits, and died in convulsions in California twelve years after the injury.

determined. There are difficulties in the way of observations upon it, which are not found in similar observations upon the cerebrum. As the result of many such observations, it has been suggested that the cerebellum in some way controls and harmonizes the voluntary muscular contractions, or, as the physiologists express it, the cerebellum is the seat of the co-ordinating power in voluntary muscular movements. This idea was at one time accepted, but it was found that in some cases of injury to the cerebellum the power of co-ordinating movements was after a time regained. And this was not the result of the healing of the injury, for the lost portion of the cerebellum is found on inspection not to have been renewed.

Diseases of this portion of the brain also give contradictory data. Tumors and injuries of the cerebellum often occasion difficulty in muscular co-ordination, but, on the other hand, no disturbance in the muscular movements has been detected in some cases, where the disease of the cerebellum has been very extensive.

Comparison of the relative development of the cerebellum in different animals seems to favor the theory proposed.

All things considered, however, scientific men are agreed that the *real function of the cerebellum is not yet definitely known.*

291. Other Ganglia of the Brain.—Besides the cerebrum and cerebellum, there are various ganglia of less size, all of which are situated on or near the under surface of the brain. Observations upon these are very difficult, and have been hitherto unproductive of results. It is an undoubted fact, however, that, unless destroyed by loss of blood, life will continue after the cerebrum and cerebellum have been removed, showing that these latter organs are not essential to life. Of the remaining ganglia, the *tuber'cula quadrigem'ina** are known to be the centers by which vision is rendered possible. The optic nerves take their origin in these small ganglia, and the fact that they receive the impressions which we call sight is proved in this way: If the optic nerve be cut between the eye and the tubercula quadrigemina, blindness results. The same effect is produced by destruction of these bodies, the optic nerves remaining uninjured. If, on the other hand, the connections between these ganglia and the cerebrum be severed, the ganglia and nerves remaining intact, sight still remains, and vision is also retained, if the cerebrum be removed and the tubercula quadrigemina left untouched. These things are sufficient to show that they are the centers of the sense of vision.

In some one or all of the remaining ganglia, with the exception of the *medulla oblongata*, to be hereafter described, the powers of sensation and motion reside. This is known from the fact that, after the removal of the cerebrum and cerebellum, the animal is still capable of sensation and voluntary motion; but, if all the ganglia be removed, excepting the medulla, not only does consciousness disappear, but all sensation and voluntary movements cease, the special senses are abolished, and the only manifestations of life are the continuance of the

---

* These small bodies are situated above and in front of the cerebellum, between it and the cerebrum, and so are not shown in any of the cuts.

functions of respiration and circulation, and the reflex movements of the parts connected with the
spinal cord.

292. The Medulla Oblongata.—The spinal cord
passes upward from the spinal canal into the skull.
Just after entering the cavity of the skull, it becomes somewhat enlarged, and this enlarged portion is called the *medulla oblongata* (Figs. 57, 58).
It is about an inch and a quarter long, and three
quarters of an inch wide, and at its upper extremity
is merged in other parts of the brain. Imbedded
in this part of the cord is a small mass of gray
matter, which is more essential for the actual preservation of life than any other portion of the brain.
It has sometimes been called the "vital knot." If
this be destroyed, all manifestations of life instantly cease. The cause of this is the intimate
relation which this part bears to the function of
respiration.

**293. The Medulla Oblongata controls Respira-
tion.**—The movements of respiration are to a certain extent under the control of the will, but, under
ordinary circumstances, we breathe unconsciously.
Inspiration and expiration take place regularly and
unceasingly, by night and by day, in our sleeping
as well as our waking hours. The nervous supply,
therefore, by which this function is controlled is,
so to speak, automatic. The requisite movements
are carried on by reflex action. How is this accomplished?

As the processes of repair and waste go on, carbon dioxide and other waste matters accumulate in
the blood. When this venous blood, loaded with
impurities, arrives at the lungs, it imparts a peculiar

stimulus to the large nerves, which are distributed to these organs. They convey a sensation to the *medulla oblongata*, and the corresponding motor nerves convey the impulse from the gray matter of the medulla oblongata, and cause a contraction of the muscular walls of the chest. Inspiration then takes place, the blood receives oxygen, the air is expired again, and the chest-walls remain passive until venous blood again accumulates in the lungs in sufficient quantity to cause the requisite stimulus in the medulla oblongata ; and so the process continues through a long life, independently of the will of the individual.

294. Automatic Action of the Medulla Oblongata. —Respiration, however, is in a measure under the control of our will. We can hold our breath, if we choose, for a short time ; but, with the cessation of respiration, the blood becomes more and more charged with impurities, until at length a peculiar sensation begins to be felt, which we call " want of breath," or "shortness of breath." If we still resist the desire to inspire, and hold the chest-walls immovable, this sensation increases until it induces the most intense suffering, and the will is at length unable to assert its power. The automatic action of the medulla oblongata overrides our feeble opposition, and we breathe again, in spite of ourselves. It is a lesser degree of this same sensation, without a doubt, which acts as a constant stimulus to the medulla oblongata, and keeps up the process of respiration without our consciousness. Now, when the portion of the medulla oblongata containing this part, called the "vital knot," is destroyed, respiration instantly ceases, and the cir-

16

culation soon stops. This is the quickest way known of killing animals, and is much used by physiologists, when they wish to produce sudden death, without injuring the blood-vessels or other organs of the body on which they wish to experiment.*

* In the great slaughtering establishments of the West, cattle are killed by puncture of the medulla oblongata. Men stand upon a raised platform, and, as the animals pass underneath, penetrate this organ with a sharp spear, or with a bullet from a carbine. The *matador*, in the Spanish bull-fights, also aims to strike this vital part, and his reputation depends upon the success with which he does it. The Spaniards execute criminals upon the same plan, by means of the *garotte*, which consists of an iron collar surrounding the throat, with a steel pin projecting inward behind. A single turn of a large wheel forces this pin into the medulla oblongata and causes instant death.

295. Nerves of the Head and Face.—In the spinal nerves, we have seen that the motor and sensory fibers are mingled, so that each nerve contains fibers of both kinds. In the nerves coming from the brain this is commonly not the case. As a rule, some nerves are entirely nerves of sensation, and others entirely nerves of motion, and the two functions are not united in a single trunk. These nerves are arranged in pairs, like the spinal nerves, one going to one side of the head or face and the other to the other.

The two most important nerves of the face are those called the *trigeminal* and the *facial*. The *trigeminal* nerve (Fig. 60) emerges from the skull by three openings, being divided into three branches just

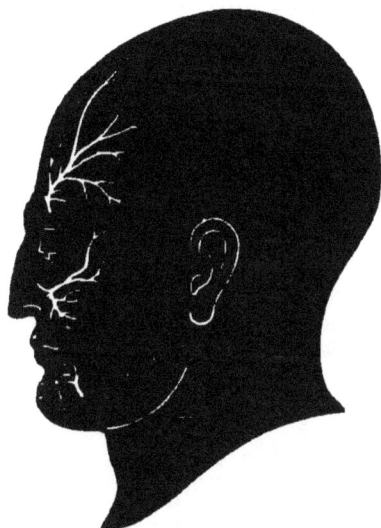

FIG. 60.—Diagram showing the distribution of the fifth nerve to the face.

before it reaches these openings. One branch supplies the parts surrounding the eye, the forehead, and the inside of the nose; the second comes out just below the eye, after sending branches to the teeth of the upper jaw, and supplies the cheek, the upper lip, and the outside of the nose; while the third comes out at a point near the front of the lower jaw, and supplies the chin and lower lip, besides having sent branches to the teeth and tongue, before emerging from the jawbone. These nerves are the great nerves of sensation of the face, and are generally considered to be the most exquisitely sensitive nerves in the whole body. They are often the seat of neuralgia, and give rise to the most intense and intolerable suffering.

The *facial* nerve (Fig. 61) is the motor nerve of the face; it emerges from the skull

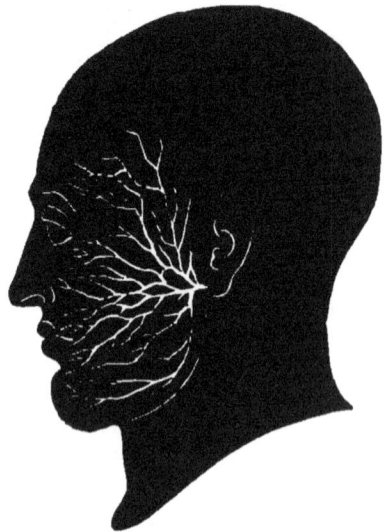

FIG. 61.—Diagram showing the distribution of the facial nerve to the face.

just below the ear, and, passing forward through the parotid gland, is distributed to all the muscles of the face. Paralysis of this nerve is far from uncommon, and produces a most singular and characteristic effect upon the countenance. Exactly one half of the face loses all expression, as much as if dead, and the contraction of the muscles of the opposite side, not being in any way counteracted, produces dreadful, though often ludicrous, distortion of the features. This form of paralysis often occurs among hackmen and others who are exposed to stormy weather. If a strong wind, particularly when accompanied by snow or sleet, be allowed to beat upon the side of the face, there is great danger of paralysis of the facial nerve. Such exposure of the place where this nerve emerges from the skull should be carefully avoided.

296. The Sciatic Nerve.—Each limb of the body is supplied by large nerve-trunks, which, as a rule, follow very nearly the course of the arteries. The largest nerve of the leg, however, runs down behind the limb, while the femoral artery runs down the front of the thigh. This nerve is known as the *sciat'ic*, and its branches extend to the foot. It is this nerve which, when pressed upon in sitting, gives rise to the sensation commonly described as the " foot being asleep."

297. Importance of Reflex Action.—It is important to call attention to the fact that *reflex action* has a great deal to do with our health and safety, even when the spinal cord and brain are in perfect condition. People in general are hardly aware how much they owe to this property of the nervous system in time of danger. Familiar illustrations are found in the rapid recovery of the bodily equilibrium, when the foot has slipped ; in the involuntary start and assumption of a position of defense, when a horse and wagon suddenly run upon one in the street ; in the start of the body at a sudden noise ; in the instantaneous withdrawal of a portion of the body when it is burned ; in the winking of the eyelids, particularly at indications of danger ;

and in hundreds of daily actions, which follow so instantaneously upon the application of a stimulus, that the thought, or mental appreciation of the stimulus, does not come until after the reflex action has already taken place.*

When we consider the immense number of such actions which take place every hour of the day, and the stupendous work of nutrition, with its complicated phenomena, all going on outside of our control, we see that in all probability the greater part of the nervous force expended in the body goes to produce involuntary movements, and to control and preserve the health and integrity of our organism in spite of ourselves. It is a suggestive thought that, much as we appear to be our own masters, independent as we seem to be of the outside world, we are really ruled by the general vital forces which equally rule all animals and even plants, upon which we are all equally dependent, and over which what we call " we " have almost as little control in ourselves as in the lion that roams the desert or in the grass that grows beneath our feet.

298. Education of the Nervous System.—The *hygiene* of the nervous system has to do mainly with the problem of *education*, and with the over- or under-use of its parts. *Education* is a broad subject, and can not be taken up here, but it is undoubtedly

---

* It is related of a distinguished chemist that a glass vessel once exploded in his hands and was blown into a thousand pieces. His hands were severely lacerated, and he at first feared that his sight was destroyed ; but on examination it was found that he had closed his eyelids involuntarily at the instant of the explosion, and, although the lids had been wounded in many places by the flying bits of glass, the eyes were uninjured. The rapidity of such reflex action far surpasses that of any voluntary movement.

true that in the future much more attention will be paid than in the past to the physical conditions which underlie all development and training of the mental powers. It is known that repetition of the same process in any portion of the nervous system renders it every time more and more easy. The result of such repetition is what we call habit. The more impressible and easily stimulated the different parts of the nervous system are, therefore, the more readily habits will be formed and the more firmly they will become fixed. Now, the cells and fibers of the nervous system of young and growing animals, like those of their muscular system, are especially soft and yielding, and, as they are constantly growing, the changes in them are more frequent, and they are much more easily influenced by any stimulus than those of adults. Consequently children and youths are especially liable to form habits, and, in order to prevent the formation of bad ones, they need the guidance and protection of older persons. The experience of the young is not wide enough to inform them correctly of what is good and what is bad, nor is it, in most cases, sufficient to enable them to appreciate the wisdom of the advice or commands of older persons. For these reasons it is necessary, in the training of the young, to require what often seems to them unreasonable, and to forbid things with what may seem to them foolish pertinacity. It is this readiness with which habits are formed in the young, and the firmness with which they become fixed, that render it so dangerous for them to indulge in sensual gratifications, such as the drinking of liquids containing alcohol, or to join in games which excite the selfish emotions in a high degree, such as lotteries and gambling.

It is this ease of habit-forming in the young, too, that renders childhood and youth the desirable time for both mental and physical training. In fact, the nervous system is, at this time, so impressionable that the training will take place inevitably, whether it is directed or not, and the success or failure of adult life is generally determined by the conditions under which the years below twenty have been passed.

299. Importance of exercising the Brain.—The nervous system, like the muscular, must be sufficiently exercised, in order to remain vigorous and healthy. If the brain be allowed to remain indolent, the faculties of memory and attention, the power of concentrating the thoughts, will be considerably weakened. It will be found that even the inactivity of a few weeks' vacation will render it a matter of some difficulty at first to keep the brain at work at one's ordinary duties. Judicious exercise also will strengthen special mental faculties, just as proper training will develop muscular power. The retentiveness of the memory may be immensely increased, and the facility with which the daily mental labor of the professional man is performed, together with the increased efficiency as age advances, shows the effect of directing the nutritive processes constantly in the same direction. In fact, this tendency is so strong that specialists, all of whose energies for many years are directed upon a particular subject, are very apt to become narrow-minded or "lop-sided," and unable to see more than one side of any subject, and that sometimes the least important one.

300. Danger of Over-Exercise.—On the other hand, too much exercise is as injurious as too little. All brain-work, as well as muscular work, involves a waste of tissue. As we are thinking, our brains are wearing out. And while this organ is actively employed, the waste of tissue is greater than the repair. When this excess of waste has reached such a point that the proper working of the brain is interfered with, we begin to feel incapable of further study, the attention begins to waver, problems seem difficult that at other times would be easy, and perhaps the memory is at fault. Such a condition demands imperative rest, and this rest is obtained by sleep.

301. Sleep.—During sleep, the repair of all the tissues goes on with great rapidity. As the body is inactive, even the brain being in sound sleep entirely unoccupied with thought, the waste of the tissues is very small, almost the whole of it being the result of the organic processes of circulation, respiration, and digestion. In all of these organic processes, which must be practically continuous, rest is obtained at short intervals, the activity of the organs concerned in them being intermittent, and, although the intervals of rest are very short, they are so numerous that they make up together about half of the twenty-four hours. But the muscular and nervous systems have no such frequent intervals of rest during waking hours, the brain, in particular, being constantly active, so that their rest has to be taken continuously in the form of sleep.

During sleep, therefore, the repair of tissue runs vastly ahead of the waste, and the exhausted parts have their vigor and irritability rapidly restored.

As the special senses and the sensitiveness of the muscular and nervous systems had become dulled and torpid because they were wasting away, so now, as they are built up again, their sensitiveness returns, and slight stimuli are sufficient to arouse them into activity. We then wake up. Even the morning light may produce a sufficient effect upon the eye, through the closed lids, to waken the sleeper, and he rises refreshed for the work of another day.

The best time for sleep is night, because it is dark and quiet; the natural stimuli of the eyes and ears are absent, and therefore rest can be obtained with the least likelihood of being broken. It is a good rule for all persons to go to bed before midnight, and rise within ten minutes after waking in the morning. The amount of sleep required by each individual must be determined by himself from experience. For most persons eight hours is not too much, and many can do with less.* Chil-

---

* Marvelous stories are told of persons who can get along with very little sleep. In most cases, if sleep is dispensed with at one time, it must be made up at another, and some people possess the faculty of falling asleep at odd moments, and making up, by what are sometimes called " cat-naps," for the loss of regular sleep. Thus, Napoleon Bonaparte is said to have been satisfied with four hours' sleep in the twenty-four; but this only refers to the time he spent in bed. He was in the habit of sleeping on horseback during a march, and his traveling-coach was made expressly so that the interior could be arranged as a bed. He often fell asleep, too, for a few minutes at a time, in his study, with his head resting on his arms on a desk in front of him. It may be doubted whether any person could live long with only four hours' sleep a day.

The Chinese sometimes punish criminals by confinement in a cage where they can neither stand up nor sit or lie down. Attendants are stationed to prevent the miserable men from falling asleep; for even under such conditions exhaustion brings on sleep. The unfortunate men are pricked and pinched, and by various means of torture kept awake until, as a rule, in eight or ten days, they die raving maniacs.

dren need more, in proportion to their fewness of years, and old people sometimes need less, because of their comparative inactivity during their waking hours.

302. Abuse of the Nervous System.—A word of caution should be spoken about the various forms of indulgence which affect the sympathetic system of nerves. These include all those forms of sensual gratification which are accompanied by a high degree of physical pleasure, followed by a more or less protracted period of depression. This depression, an instance of which is seen in the reaction following intoxication, is the result of a great shock and consequent great waste of tissue in the sympathetic nervous system. Recovery from such shocks is very slow. The sympathetic ganglia have to do with the important function of nutrition. They lie at the very foundation of life in our bodies. Their ordinary duties are severe, and about all they can attend to properly. Therefore, any great shock to this part of the nervous system affects the very centers of life. This is the reason of the "breaking up," as it is called, which results from the various forms of dissipation. This word "dissipation" is well used for such acts as are commonly included under this head, as all kinds of debauchery, abuse of the digestive organs, late hours, insufficient sleep and relaxation, etc., dissipate nervous force with frightful rapidity.

303. Effects of Alcohol upon the Nervous System.—Of all the means by which the nervous system is abused alcohol probably stands first, by reason of its extensive use and the very marked deleterious effect it is likely to produce. Immediately upon the introduction of an alcoholic drink into the stomach a sensation of warmth is perceived, followed by slight exhilaration. This is probably due to the irritation of the nerves of the stomach, even before the alcohol is absorbed. When the alcohol has been taken into the blood and reaches all parts of the body its peculiar effects upon the nervous system become apparent. There is at first a stage of excitement, when the brain seems to be more active, ideas flow freely, there is a disposition to laugh at trifles, and speech is less circumspect than natural. Even in this stage of intoxication, however, when the activity of the brain seems to be increased, a close and sober observer can see that a poisonous effect has been produced upon the nervous system. He can detect a loss of control, a slight diminution of will-power, a slight lack of coherence in trains of thought, and, on the other hand, an increased tendency to laugh, to declaim, to joke, to banter, which shows a diminution of intellectual power and an increased sensitiveness of the emotional nature, the latter being, however, not an actual increase, but a relative one, the balance of the mind being disturbed and the dominance of the intellect over the emotions being partially suspended. The judgment is weakened, control over the thoughts is lessened, and things which when strictly sober the drinker conceals often come to light in his conversation. In brief, he is beginning to lose the mastery over himself.

304. **Effect of Larger Doses of Alcohol.**—If more alcohol is taken this diminution of the regulative force of the intellect becomes more marked. The emotions have free sway. The person becomes

maudlin, he laughs or cries with equal readiness, becomes ludicrously affectionate or causelessly suspicious, he is obsequious in his politeness or obstinate to rudeness, and his intellect is so far in abeyance that his judgment is lost, he can no longer weigh consequences or reason upon occurrences, and is the prey of every suggestion. The movements of the voluntary muscles are no longer under control. The currents of force sent along the nerves are irregular and inco-ordinate. The tongue seems thick and unmanageable, and speech is mumbling and indistinct; the lower limbs are moved awkwardly and with a staggering gait; the muscles of the eyeballs no longer act in unison, but the axes of the eyes are guided in different directions, and double vision is the result. All of the senses are dulled, and the person often suffers injuries, of which at the time he is unconscious. In this self-inflicted degradation the drinker reduces himself below the level of the lowest brute. But it should be remembered that the first glass is the first step toward such a state.

305. Effects of a Very Large Dose of Alcohol. —If still more alcohol is taken, this condition rapidly merges into one of complete insensibility, and the drunken man lies like a log, with muscles relaxed, with puffing or snoring respiration, with glassy eyes, and, in extreme cases, abolition of all functions, excepting those of vegetative life. Such a condition may lapse into death, but more frequently the alcohol is gradually eliminated from the body by the lungs, kidneys, and skin, and recovery takes place.

Such a debauch is followed by extreme nervous

prostration, with great lassitude, dull headache, giddiness, and sometimes dimness and yellowness of vision on stooping and suddenly rising again, accompanied by the disturbances of the digestive organs already described. These symptoms are surely not the effects of a food but of a rank poison.

306. How these Effects of Alcohol are Produced.—It is believed that alcohol enters into actual combination with nervous tissue, and produces the effects just described by preventing the normal nutritive changes. This effect manifests itself as a deadening of nervous sensibility, usually termed narcosis.

307. Chronic Alcoholism.—When alcohol is habitually taken in sufficient quantities, what is called chronic alcoholism is the result. The symptoms of this condition are mainly due to the effect of alcohol upon the nervous system. The first indication is generally muscular tremor, especially on rising in the morning. At first this tremor may be stopped by a special effort of the will, but it soon becomes uncontrollable. A glass of spirits, however, will steady the muscles, and old topers think they must always have this early in the morning. Soon the sleep begins to be disturbed. The person, even if drowsy when he goes to bed, tosses restlessly from side to side, unable to remain long in one position, not from any pain or discomfort, but simply from an indescribable feeling of disquietude. If he obtains broken snatches of sleep, he has frightful dreams. He rises unrefreshed, and, even if he attends to his daily occupation, is harassed by a constant feeling of dread and anxiety for which he can assign no cause. He is troubled with noises in the ears and headache. The mind becomes seriously impaired, the will is weak, and great uncertainty of purpose is noticed by friends, and perhaps by the sufferer himself. When the chronic poisoning has reached this stage the drunkard is on the verge of delirium tremens.

308. Delirium Tremens.—This frightful disease may be the result of a single prolonged debauch, but more commonly supervenes upon a long course of drunkenness, or upon a greater indulgence than common in persons who call themselves moderate drinkers, i. e., persons who daily drink a considerable amount of alcoholic liquor, but who never become drunk. It may come on during a debauch or after its close. The first symptom is absolute sleeplessness. The sufferer is restless and troubled with terrifying visions, usually of disgusting or horrible objects, black and crawling. He sees insects, reptiles, or vermin on the floor or creeping over the bed. These delusions persist in the daytime, and he is constantly occupied with trying to escape, pushing them away or resisting them with violence. He often hears voices close to his ear, mocking or threatening him. He is unable to take food, and the muscular tremors previously spoken of are incessant and distressing. In this pitiable condition he may remain for several days, a sight of horror, until sleep or death comes to his relief. In the one case he generally wakes convalescent, though greatly prostrated. In the other he often dies in the midst of the most frightful visions, expending the feeble remnant of his strength fighting his imaginary foes.

Chronic alcoholism sometimes produces attacks

of insanity, in which the unhappy man tries to kill himself or others.

309. Moral Deterioration due to Alcohol.—A striking effect of alcoholic habits is the moral deterioration they produce. The drinker at first loses the finer and nobler traits of character, and gradually, as the habit increases, becomes untruthful, dishonest, and lazy; he loses his self-respect, and becomes so shameless and regardless of the laws of society and the state that he often has to be cast off and disowned by his dearest friends for their own protection. Many paupers owe their degradation to this alone. It should be borne in mind that it is the nature of alcohol, though taken only in small quantities, to create the appetite that leads to these results.

310. Effects of Alcohol on Nervous Tissue.—It has been found that the nervous system of drunkards undergoes marked changes, which can be detected by the microscope. They are of a similar nature to those occurring in other parts of the body, viz., an increase of connective tissue and of fatty matters. The nerve cells and fibers become shrunken and wasted; there may be softening of the brain, or small hæmorrhages here and there upon its surface; the enveloping membranes of the brain are thickened, and the whole amount of the brain-substance is lessened, the cavity of the skull being kept full by the transudation of serum from the blood.

311. Hereditary Effects of Alcohol.—The degenerative changes produced by alcohol upon the nervous system are not confined to the drinker alone, but may reappear in his children and grandchildren. These may manifest themselves in various ways, either by the recurrence of the alcoholic appetite in the child, by tendencies to immorality, by nervous disorders like epilepsy, idiocy, and insanity, or by a weak condition of the nervous system in which the more serious disorders may result from slight causes. Instances have been known where all the children of a family born after the father began to drink were either weak-minded or had some serious nervous disorder.

312. Other Poisonous Substances in Alcoholic Drinks.—The bad effects which have been described as due to the use of beverages containing alcohol are increased by the fusel-oil and ethers contained in them. All liquors distilled from grain, potatoes, or other starchy substances, contain fusel-oil (or, as it is also called, amylic alcohol). This is especially injurious, and delirium tremens and other serious derangements of the nervous system occur much more frequently in persons who drink liquors containing it—such as new whisky, and gins, brandies, or wines, which have been manufactured out of potato-spirit, or to which it has been added. The oil of worm-wood, which is found in absinthe, is exceedingly dangerous, and rapidly produces in those who use it a peculiar form of insanity with muscular tremors, giddiness, and epileptic convulsions. These oils also develop the various degenerations of organs much more rapidly than ethylic or common alcohol.

313. Effect of Tobacco on the Nervous System. —The great and general depression produced by tobacco, when used for the first time, is mainly due to its effect upon the nervous system. The feeble-

17

ness of the heart's action, the pallor of the surface, the distress at the pit of the stomach, are all due to a poisoning of great nerve-centers by the nicotine. But, as with alcohol and opium, the nervous system gradually accommodates itself, after a fashion, to repeated attacks of this poison, so that its depressing effect is only felt in a slight degree when the weed is used in moderate amount. It then dulls the sensibilities, so that the man who is tired and needs rest does not realize his weariness. It is probable that this effect is produced in the same way and for the same reason as the nausea and giddiness that mark the first experience with tobacco. It is due to a diminution of the blood-supply in the brain. The brain is pallid, if we could see it, like the skin. Now, when the mind is worried or irritated, the brain is congested, the blood-vessels are overfilled, and the tissue-changes are more rapid than is healthy. The proper and natural relief in such cases comes from rest and the withdrawal of all cause of irritation, when, in a healthy person, Nature will restore the circulation to its normal condition. But man knows so much better how to manage himself that he introduces a poison into his body to depress the heart's action and render it too weak to keep the small blood-vessels well filled. As the blood-supply is lessened the changes in the brain-cells are less rapid, and the person calls himself soothed, when he is, in fact, poisoned.

It is not known that tobacco causes permanent injury to the structure of the brain, as alcohol does, but it will sometimes give rise to wakefulness, interfering greatly with rest at night, making the smoker nervous and irritable, with tremulous mus-

cles, thus bringing about apparently the very condition it is intended to control. This form of
nervous irritability, however, is due to a failure of
proper nutrition of the brain, by reason of a lack
in blood-supply, while the other form is due to
oversensitiveness of the nerve-cells, caused by an
excess of blood in the brain. When tobacco is
abandoned, however, the bad symptoms disappear
with signal rapidity.

**314. General Effect of Opium on the Nervous
System.**—Opium produces its greatest effect upon
the nervous system. Unlike alcohol, its use is not
followed by extensive degeneration of organs and
tissues, so far as can be discerned by examination
after death. It must, however, produce a lasting
alteration of the nerve-cells, or it would not give
rise to the great psychical changes which invariably attend its habitual use.

315. Effect of a Small Dose of Opium.—A small
dose of opium—one grain, for instance—is followed
in a short time by moderate excitement, the pulse
growing stronger and eight or ten beats quicker to
the minute. It falls again, however, in half or three
quarters of an hour. The mouth and throat are
dry, and perspiration often breaks out. The brain
and nervous system soon feel the narcotic effect of
the drug, and in forty or fifty minutes the person
grows sleepy, and, if it be night, will probably sleep
all night, though in the daytime he will wake, perhaps in two or three hours. After waking there is
some depression and disturbance of the digestive
organs.

316. Effect of a Larger Dose of Opium.—If the
dose is larger—two or three grains—it is followed

by great excitement, with noises in the ears, and the ideas are confused and extravagant. The head soon feels heavy and dull, the senses are blunted, and deep sleep follows, attended by dreams of a disagreeable nature. The reaction after narcotism has passed off is proportionately greater.

317. Effect of a Dangerously Large Dose.— After a large dose, one which involves danger of death, the stage of excitement is very brief, there is a sense of giddiness and oppression, soon followed by profound sleep. The pulse, which at first was full and strong, becomes small and feeble. The pupils are strongly contracted, often like a pinpoint in size. The face is pale, the respiration slow and labored, and the muscles are in a state of complete relaxation. Such a condition soon passes into death, sometimes preceded by convulsions.

318. The Opium - Habit.—The sensations produced by a small dose of opium in some persons tempt them strongly to take it again. It is at first a luxury, a mere matter of sensual gratification, but, before the individual is aware of the fact, he is the slave of the opium-habit. He finds that the agreeable feelings induced by the opium he has been taking begin to be replaced by others of a horrible or disgusting character. After the effects of a dose have passed away he is gloomy, depressed, and wretched, and feels a ravenous gnawing of the stomach, which can only be appeased by more opium. Then he begins to struggle against his chains, and the more hopeless he finds the struggle the deeper becomes his despair and gloom. The drug no longer affords him pleasure ; it merely relieves his suffering. He takes it now not for enjoyment, but to escape unspeakable misery, and to his horror finds that he must constantly increase the dose.

319. Mental, Moral, and Physical Deterioration produced by Opium.—The moral deterioration resulting from the habitual use of opium is fully equal to that caused by alcohol. The opium-eater becomes untruthful, slothful, unambitious, and careless, wrapped up in the gratification of his unnatural appetite, and regardless of the means he adopts to satisfy it.

The memory fails, the temper becomes fanciful and capricious, the will is broken, energy disappears, giddiness, neuralgias, and sleeplessness supervene, and partial paralysis of the muscles of the back and limbs is common, so that the sufferer shuffles along bent, shriveled, and weak, like a decrepit old man.

Death often results from marasmus, or a gradual wasting away, due to the general failure of nutrition, the effect of the drug upon the digestive apparatus.

320. Effects of the Chloroform-Habit.—Chloroform is eliminated so rapidly by the lungs that the period of excitement is very short, and its habitual use as an intoxicant is for its stupefying effect. For some persons the gradual lapse into insensibility is attended by such pleasurable sensations that they acquire the habit of inhaling chloroform to produce it. Its long-continued use is followed by sleeplessness, tremulousness of the muscles, and marked inability to fix the attention upon any subject. If its use is persisted in after these preliminary symptoms have appeared, it retards the tissue changes in the blood, and produces marasmus and a melancholic form of insanity.   The drug is a very dangerous one, tending to produce sudden death by paralyzing the nervous ganglia of the heart.

321. Effects of the Chloral-Habit.—Chloral, in moderate doses, fifteen or twenty grains, brings on a sleep very much like the natural sleep.   The person can be easily aroused from it, and it is followed after waking by no depression or nausea.   It thus becomes a valuable drug in the hands of physicians.   If it is habitually used, however, it is followed by various disagreeable effects.   The blood becomes thin and watery, and eruptions or blotches occur on the skin; there is some irritation of the eyes and throat, shortness of breath, and pains in the limbs.   There may be even partial paralysis of the lower limbs, and there is great deterioration of the mental and moral nature.   Death may result from gradual failure of the heart, or more commonly from the effect of a poisonous dose inadvertently taken.

322. Cause of the Craving for Narcotics.—The unnatural craving felt by the drinker is shared by all who become slaves to a narcotic.   It is probably due to the fact that these poisons enter into actual contact with the nerve substance.   This has been shown to be true of alcohol, and although the changes produced by the other narcotics are not known, the similarity of their effect upon the nervous system suggests that they may act in the same way.   A combination between the alcohol and the nerve-matter of cells and fibers renders the composition of the nervous substance different from normal, but the body accustoms itself, after a while, to the new order of things.   When the alcohol has disappeared from the tissues of the body the nerve-cells are left in a state of disorganization. Disorganization of cell-structure means the beginning of death, and the whole body sends up a cry for relief. And the relief must be immediate and easily attained in order to abate suffering. If the drunkard whose nervous system has been brought to this state, in which the cells are in a continual state of unrest unless combined with alcohol, can be prevented from obtaining it for a considerable time, the cells will gradually become able to appropriate food material they need from the blood, and, after a time, resume their normal condition. Then the craving ceases.

This craving for narcotics is the most terrible fact connected with their use. It constitutes a slavery of the most dreadful kind. It is more hopeless in some cases than in others. The chains of the alcohol slavery may be broken. The use of tobacco may perhaps be given up more easily. But the slavery to opium is pitifully lasting. Few ever escape from its toils. It is even doubtful if any do, when the net is once fairly cast about them.

PART V.

ORGANS OF PROTECTION.

CHAPTER I.

THE SKIN.

323. The Skin.—The skin, with its appendages, is a complex organ, and has many functions. It serves for protection, contains nerves of sensation, secretes, absorbs, and excretes, and its functions are so necessary to life that, if in any way it be rendered unable to perform its duty in the organism, death is the result.

324. Structure of the Skin.—The skin is composed essentially of two layers, one called the *cuticle* or *epidermis*, and the other the *cutis* or *derma* (Fig. 62).* The former is the outside layer, and the latter the inside. The *epidermis* consists of small, flattened, dry scales, and receives the brunt of injuries and abrasions. The thickness of this layer is shown in a blister, the epidermis being there separated from the layer beneath by the fluid which fills the vesicle.† The inner layer, or *derma*, constitutes

---

* *Cutis* is the Latin for *skin*, and *cuticle* means *little skin ; derma* is the Greek for *skin*, and *epidermis* means the *outer layer of the skin*, or, literally, "*upon the skin.*"

† *Vesicle* is from the Latin, and means *a bladder*, which a blister somewhat resembles.

what is sometimes called the true skin. This layer alone contains nerves, blood-vessels, hair-bulbs, and

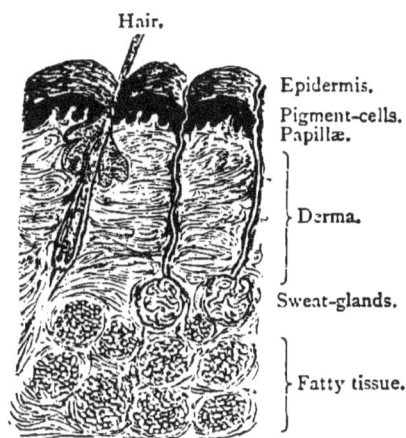

FIG. 62.—Diagram representing a vertical section of the skin. Attached to the hair are two sebaceous glands.

the other appendages of the skin. It is made up of strong interlacing fibers of connective tissue, which form a firm, close covering for the delicate tissues beneath. This layer varies in thickness in different parts of the body, from one fiftieth to about one sixth of an inch. For instance, the skin of the eyelid is very thin, while that of the small of the back is the thickest in the body.

325. The Derma. — The under surface of the derma is merged in the loose connective tissue which lies between the skin and the flesh. It is in this loose tissue that fat accumulates, and, in consequence of its peculiarly loose and flexible structure, the skin can be moved easily in any direction, slipping or gliding, with no pain and no disagreeable sensation, over the tissues immediately beneath it. The upper surface of the derma is covered with

little elevations, called *papil'læ*, which, on an aver-
age, are about $\frac{1}{100}$ of an inch in length, and $\frac{1}{250}$ of
an inch broad at the bottom. They are somewhat
conical in shape, and in their interior are found the
ends of sensitive nerves. In some parts of the
body they are scattered irregularly on the surface
of the derma, while in other regions they are arranged in rows, and produce the appearance of
small parallel ridges on the surface of the body.
This latter arrangement is most distinctly to be
seen on the palm of the hand and fingers, and in
these parts also the sense of touch is most highly
developed.

326. The Epidermis.—The *epidermis* lies in immediate contact with the derma, its under surface
being accurately molded to fit the papillæ. But, in
the under portion of it, the portion into which the
papillæ project, the scales, instead of being flat and
hard, are rounded and soft, and contain a certain
amount of coloring-matter, in the form of small,
dark granules. As this part of the epidermis
varies in color and thickness, so the complexion
of the person varies.* In the negro, these pigment-cells, as they are called, are very dark and
numerous, and there are all gradations in the
amount of the coloring-matter in different persons,
from the coal-black African to the albino, in whom
there is no color at all, except the red tinge given
by the blood.

* This layer of pigment-cells seems to protect the parts beneath
from the action of the sun. The natives of all tropical countries are
dark, and exposure to the sun's rays, even in our climate, tans the skin
to a darker hue. Sometimes this effect of the sun is more noticeable
in particular spots, producing what are known as freckles.

327. The Nails.—Near the ends of the fingers and toes the skin takes on a peculiar appearance. The epidermis is very closely applied to the derma, with no colored cells between them, and the scales of which it is composed are singularly dry and hard, and semi-transparent. These peculiar growths are called the *nails*. The part of the finger from which they grow is called the root of the nail, and the nail grows both in length and thickness. It is formed by the continual pushing forward of new cells, both from the root and from the under surface. The growth from the root makes it grow in length, while the addition of cells on the under surface renders it thicker at the forward end than it is near the root. The nails were probably once a means of defense and offense, when men were in a low, barbarian state; but, at the present day, their chief value appears to be for ornament, and as a support for the pulp of the finger.*

328. The Hair.—The *hairs* are peculiar growths of the skin. At certain points the surface of the skin seems to dip inward, forming a deep pit or pocket, at the bottom of which is a little projection or papilla. From this papilla the hair grows, and, emerging from the pocket, attains in some parts of the body a considerable length.† The body, or shaft, of a hair consists of cells, flattened and hard, but through its center, from one end to the other,

* The nails are of special use in the way last mentioned. The finger-tips are extremely sensitive, and the hard, stiff nail serves as a backing against which it can be pressed by the objects it touches. In this way the nails increase materially the delicacy of touch.

† Instances have been known of women with hair six or seven feet in length, and of men whose beards measured nearly as much. Such cases, however, are very rare.

runs a hollow canal, filled with pigment-cells. These cells, according to the amount of the coloring-matter in them, give the hair the appearance known as black, brown, gray, auburn, etc. Almost the whole surface of the body is covered with hair, excepting the palms of the hands and the soles of the feet, but most of it is of a soft, downy variety, which is not visible without close observation.

329. The Sebaceous Glands.—There are certain small glands, situated in the substance of the derma, which are called the *sebaceous* * *glands*, or sometimes the oil-glands. They are composed of minute sacs lined with epithelial cells, which constantly secrete and pour out of their mouths a whitish, tallow-like, fatty substance. Most of them empty into the hair-pits or follicles,† as they are called, and thence their secretion runs along the hair, and out on the surface of the body. They aid in keeping the hair and skin soft and flexible. Sometimes the opening, by which the secretion of the gland is discharged, becomes obstructed, and the gland still goes on secreting. In such a case, a swelling makes its appearance at the seat of the trouble, and the irritation of the retained secretion causes an inflammation to set in, which continues until the contents of the inflamed sac are discharged, or let out by the prick of a needle. This is the cause of the unsightly little pimples or boils which occasionally come on the end of the nose, and give rise to great annoyance.‡

---

* *Seba'ceous*, from the Latin *sebum*, meaning *fat*, or tallow.

† *Fol'licle*, from the Latin word *folliculus*, meaning a little *sack* or *pouch*.

‡ There are many of these sebaceous glands about the end of the nose, and the little openings, by which their secretion reaches the surface, often become obstructed by dirt, so that the secretion accumulates

330. The Sweat-Glands.—There are still other glands in the skin which have a very important part to play in the maintenance of life itself. These are the *sweat-glands*, or *perspiratory glands.* They are situated either in the lowest layer of the derma or in the loose connective tissue just beneath it. The mass of one of these glands is convoluted— i. e., it is made up of a long tube, rolled about and twisted into a ball. This tube is lined with epithelial cells, like the interior of all glands, and, after leaving the convoluted portion, it passes in quite a direct course through the derma, and then in a spiral course, like the thread of a corkscrew, through the epidermis, on the surface of which it opens. If one of these little glands be unrolled, and its tube stretched out, it is found to be about $\frac{1}{15}$ of an inch in length, while its diameter is about $\frac{1}{400}$ of an inch. Notwithstanding their minute size, however, there are so many of them that their secreting surface is enormous. They are least numerous on the neck and back, where there are 417 to the square inch, and they are found in the greatest numbers on the palm of the hand, where there are no less than 2,700 of them to the square inch. The total number is said not to be less than 2,300,-000, and by a simple calculation it will be found that the total length of these minute tubes, if carefully unrolled and placed end to end, would be about two miles and a half!

in the interior. When this is squeezed out, it looks like a small, white worm, the dirt at the outer end resembling the head. Hence the term "black-heads." To be sure, there are sometimes little worms found imbedded in this secretion, but they are only to be seen with the microscope.

331. Effects of Pressure on the Skin.—It is evident, without demonstration, that the skin protects the parts beneath it. In the performance of this function, it sometimes resents too great pressure, or unequal pressure, from outside. In such cases the skin becomes thicker in spots where the pressure occurs. If the epidermis only is affected in such cases, and there is merely an increased production of epidermal cells from beneath, the skin becomes, as we say, *callous*. If, however, the derma be affected, the papillæ become enlarged and project above the surface, *warts* are produced, and, where both forms of growth take place, we have *corns*. Pressure is undoubtedly not the only cause of the development of these unpleasant diseases, but it is one cause. But the origin of warts, in particular, is generally obscure.

332. Animal Heat.—The skin also has an important function to perform in maintaining a uniform temperature in the body. Every one knows that a living body is warm. Why is it so? There has been much discussion on this point, and many theories proposed regarding it. When oxygen was first discovered, Lavoisier* supposed that the process of respiration was a process of combustion; that the oxygen which entered the lungs in the air united directly with the carbon in the venous blood, and produced carbon dioxide. In ordinary flame, this is the chemical change which accompanies combustion. The oxygen of the air unites with the carbon of the tallow or oil and produces carbon dioxide, the hydrogen of the oil also uniting with oxygen to produce water. Thus we get, as a result of combustion, a disappearance of oxygen and a production of carbon dioxide and water. Now, in the air which passes in and out of the lungs, we find the same thing. Oxygen passes in, and carbon dioxide and water, in the form of vapor, come out. What more natural, then, than to conclude that respiration was combustion, and the lungs a sort of stove by which the heat of the body was kept up?

It was soon found, however, that the lungs were no warmer than the rest of the body, and particularly that the blood contained carbon dioxide before it reached the lungs, and took away free oxygen from them. These facts showed conclusively that the change did not take place in those organs.

The great chemist, Liebig,† then changed the the-

---

* Antoine Laurent Lavoisier, a celebrated French chemist (1743–1794). He first discovered the composition of water, and invented a chemical nomenclature which lasted for more than fifty years. He was guillotined during the French Revolution, because, twenty years before, he had been a tax-gatherer under the monarchy.

† Baron Justus von Liebig, a renowned German chemist (1803–1873), Professor of Chemistry at Munich. He was the greatest chemist of his time, made many practical applications of chemistry in agriculture and the arts, and was one of the founders of the science of organic chemistry.

ory, by supposing that the combustion took place, not in the lungs, but in the tissues generally through-out the body, and this theory was for a long time accepted. It is now giving way, however, as it has been found that many other chemical processes, besides the union of carbon and oxygen, give forth heat, and that the amount of oxygen inhaled with the breath hardly ever corresponds with the amount contained in the carbon dioxide expired. It is now considered that the heat of the body is the result, not of one class of chemical phenomena, but of all the molecular changes and combinations which are constantly taking place in all parts of the body.

333. Normal Temperature of the Human Body. —The temperature natural to any animal is found to be such that, if the blood be made very much cooler or very much warmer, it is fatal to life. The proper temperature of the interior of the human body is about 100° Fahr. On the surface, of course, some parts are cooler than others, according to the degree of exposure to which they have been subjected. Physicians usually take the temperature of patients by placing the bulb of a thermometer in the armpit, and the instrument in this spot generally registers about 98½° Fahr. The normal temperature often varies ½° or 1° from this standard, according to the time of day or night, and according to the sleeping or waking condition of the person.

334. Great Variations in External Temperature. —Now, if the blood be cooled much below this point, say down to 80°, death takes place; equally so if it be heated to a temperature of 113°. In the former case, the person is said to have been frozen, for the temperature can not be reduced to this point unless by exposure to a very frigid atmosphere; and, in the latter case, he is said to die of sunstroke, because such a temperature is rarely induced, excepting in extremely hot weather.

But persons are often exposed to a temperature of 10° or 20° below zero, and of from 100° to 130° above zero. The latter figures, perhaps, require a word of explanation. We are apt to think of the temperature to which we are exposed in a hot summer's day as the same which we find registered by a thermometer hanging in a sheltered spot. This is an incorrect way to estimate it. Many of us, perhaps most of us, spend some time every day in the direct rays of the sun, and it is no uncommon thing for a thermometer placed in the same situation to register 130°. How, then, are we enabled to resist such extremes of cold and heat, and maintain our bodily temperature almost unchanged?

335. Effect of Cold on the Body.—When the body is exposed to a low temperature, and the surface is being continually cooled by radiation, the first effect, as the cold reaches the tissues just under the skin, appears to be a stimulating one; the processes of nutrition, of cell destruction and formation, take place with greater rapidity, and therefore more food is required to supply the waste. We all know that in cold weather our appetites are more keen, and, if we digest more, we must waste more, or else accumulate material in our bodies. This latter phenomenon does take place to some extent, and most persons gain flesh during cold weather.* Now, these changes in nutrition, with

* Cold weather produces a craving for fatty foods. The Esquimaux eat enormous quantities of blubber. Dr. Hayes, the Arctic explorer,

18

increased activity, give rise to an increase of temperature, and, up to a certain point of exposure, this increase of heat is sufficient to balance the increased radiation. When the radiation becomes excessive, and the greatest activity of the bodily functions is insufficient to produce enough heat to compensate for it, the blood begins to grow cool, and the symptoms of drowsiness and stupor come on, which are the precursors of death.

336. Effect of Heat on the Body.—On the other hand, when the temperature is high, the tendency to the accumulation of too much heat in the body is kept down partly by a diminution of activity in the nutritive processes, and partly by the action of the perspiratory glands. In hot weather, the nutritive processes go on much more slowly than in cold, we eat much less food, most of us lose flesh, and the production of animal heat in the interior of the body is thus very much reduced. But this reduction is not sufficient to guard us against the harmful effects of a very high temperature. If cold becomes too great to be resisted by the ordinary forces at work within the body, we put on additional clothing, build ourselves fires, shelter ourselves in houses, etc., and thus keep warm. But in hot weather, if we had no perspiratory glands, we should be badly off indeed.

337. How the Temperature of the Body is regulated.—The way in which these glands regulate the temperature of our bodies is by covering the sur-

states that his men often drank clear oil with great relish, and the Laplanders are said to be fond of tallow-candles. Even in our own climate, fried articles of food, which contain a good deal of fat, are palatable in winter, but very distasteful in warm weather.

face of the skin with a watery fluid, whose evaporation continually abstracts heat. Evaporation consists in the change of a liquid into vapor. The evaporation of any liquid causes a diminution of temperature in whatever lies in contact with it. Liquids which evaporate very rapidly, like ether, produce a very striking sensation of cold, and may even be used, in the form of spray, to freeze a portion of the body. Now when the surface of the body is heated, the vaso-motor nerves are somewhat paralyzed, and an unusual flow of blood takes place to the capillaries of the skin. A great flow of blood to a gland causes an increased secretion, and accordingly the perspiratory glands immediately begin to pour forth their peculiar secretion. The perspiration is composed of water and salts, containing about $99\frac{1}{2}$ per cent of water. This water evaporates rapidly, and abstracts so much heat that the temperature of the body is kept at its normal standard.

338. Effect of Alcohol on the Temperature of the Body.—Experiments to determine the effect of alcohol upon the bodily temperature have given very contradictory results. In healthy persons, unaccustomed to its use, full doses of alcohol produce a slight fall of temperature, while if the quantity is large enough to intoxicate there may be a fall of 3° or 4° Fahr., lasting for several hours. In persons who are in the habit of drinking the temperature will often not be affected even by quite large doses.

It was formerly believed that the use of alcoholic liquors enabled men to endure extremes of heat and cold better than they could without such assistance; but the testimony of travelers, explorers, and army officers is now overwhelmingly against any such proposition. The action of alcohol in paralyzing temporarily the vaso-motor nerves results in the overfilling of the capillary vessels and a great flow of blood to the surface of the body. This gives rise to a feeling of warmth. Every one knows that in blushing the face is not only red but hot. But this very flow of blood to the surface results in an increased radiation of heat, and the temperature of the blood becomes lowered. Thus the feeling of warmth produced by a dose of alcohol is a deceptive one, for it really lowers the temperature of the body, and thus renders it less able to resist external cold. The officers of Arctic expeditions are unanimous in their opinion that the use of alcohol in cold climates is injurious.

In hot climates, on the other hand, there is less radiation from the body and less food is required to keep up the animal heat. The tissue changes are less active than they should be for perfect health, and, as alcohol tends to prevent such changes, it increases the pernicious effects of the high temperature. In army marches in tropical climates it has been found that the drinkers of alcohol do not bear the heat so well as the abstinent, and that they are more liable to sunstroke.*

Some striking instances have been given by Dr. Parkes, and others, which illustrate these statements. The Russian soldiers, for example, when marching in winter use no spirits, and, what is

---

* It has also been shown that hard labor, either physical or mental, is better done without alcohol than with it. This fact is recognized and acted upon in the best training of athletes for races and games, where such drinks are not allowed.

more, no man who has lately taken any is allowed to start. The Swiss guides in Chamonix and the Oberland are very sparing in the use of spirits in cold weather, finding it invariably hurtful.

In India, Annesley, after twenty years' service, declared that dram-drinking killed more persons than the climate or the sword. In the Kaffir war, at the Cape of Good Hope, in 1852, "a march was made by two hundred men from Graham's Town to Bloemfontein and back; one thousand miles were covered in seventy-one days; the men were almost naked, were exposed to great variations of temperature (excessive heat during the day, while at night water froze in a tent where twenty-one men were sleeping), and got as rations only biscuit, half a pound of meat, and what game they could kill. For drink they had nothing but water. Yet no man was sick until the end of the march, when two men got dysentery, and these were the only two who had the chance of getting any liquor."

339. Amount of Perspiration.—The secretion of the perspiration takes place constantly. Under ordinary circumstances, its amount is so small that it does not collect in drops, but evaporates as soon as it reaches the surface. This is called the " insensible perspiration," and takes place at all times. Although so gradually secreted and so quickly evaporated as to be unperceived by the individual, its daily amount is surprisingly large. Lavoisier and Seguin have found it to be a little less than two pounds. This amount is immensely increased when the body is exposed to an elevated temperature, and the perspiration begins to run from the pores in streams. It then often rises to the amount of a pound an hour, and, as this great loss of fluid from the body has to be resupplied from without, such excessive perspiration is followed by excessive thirst.

340. Exposure to Dry Heat.—In order for evaporation to take place rapidly, the air must be dry enough to absorb the vapor into which the water passes. If the air be very dry, and contain very little water, so that it is prepared to absorb an immense amount of it, the human body can endure without injury an incredibly high temperature. Human beings have remained in ovens heated to a temperature of from 350° to 600° Fahr., and remained there while eggs and even beefsteaks were cooked by their side.*

341. Exposure to Moist Heat.—If, on the other hand, the air be already so saturated with watery vapor that it can not take up much more (for its capacity in this respect is limited), any great rise in temperature causes extreme discomfort. It is on account of this difference in the moisture of the atmosphere, and consequent diminution or increase of evaporation, that our physical comfort varies so much on different days, when the thermometer registers the same temperature. For this reason, also (the moister atmosphere), a temperature of 85° is almost unendurable in London, while in New York it is usually borne without discomfort.

---

* " The workmen of the sculptor Chantrey were in the habit, according to Dr. Carpenter, of entering a furnace in which the air was heated up to 350° " (Dalton). A public performer, named Chabert, who called himself the " Fire-King," is said to have exposed himself to a temperature of 600°, and remained in an oven heated to that degree while a beefsteak was cooked by his side. In such experiments care has to be taken that no metal, or other good conductor of heat, comes in contact with the body, for if it does it will cause frightful burns.

342. Respiration and Absorption through the Skin.

—There is also a certain amount of what might be called respiration going on through the skin—i. e., it has been shown, by inclosing one of the limbs in an air-tight case, that the contained air loses oxygen and gains carbon dioxide, showing that an interchange of those gases takes place through the skin, as well as through the lungs.* It is also a fact, well known to physicians, that the skin possesses the property of absorbing various substances placed in contact with it.†

The skin is thus seen to be an exceedingly complex organ, and to possess many very important functions. There are certain peculiarities attending injuries and diseases of the skin which have led many physiologists to think it has important functions not yet discovered.‡

The *hygiene* of the skin has mainly to do with the questions of how to keep it cleansed from impurities, and how to regulate its temperature—i. e., with *bathing* and *clothing*.

---

* The amount of carbon dioxide thrown off by the skin is estimated at about one thirtieth of that exhaled from the lungs.

† Medicines are sometimes administered by rubbing them into the skin. Castor-oil rubbed over the stomach will produce a medicinal effect, and so will mercury. Infants who were too much exhausted by disease to eat, or whose stomachs refused to retain food, have been saved from death by rubbing nutritive substances, oils, etc., upon the surface of the body.

‡ When the surface of the body is completely covered with a coating impervious to air, death ensues very rapidly. At the coronation of Giovanni de' Medici (1475–1521) as Pope Leo X, a little boy was covered with gold-foil to represent a cherub, and add to the splendor of the ceremonies. He became almost immediately ill, however, and, in spite of all that was done, died in a few hours, because the gold and varnish were not removed.

343. Necessity of cleansing the Skin.—The impurities on the surface of the skin come mainly from three sources: 1. The perspiration contains a small amount of solid matter, mostly mineral, which is deposited upon the skin and left there as the water evaporates. 2. The fatty secretion of the sebaceous glands is constantly being poured out in small quantity, and dries upon the surface. 3. The epidermal scales in the outer layer of the skin are being continually shed, pushed off, as it were, by the cells developing under them, and most of them fall away in the form of a fine, branny dust, which clings to the under-clothing, and is brushed and washed from other parts of the body. A certain amount of this refuse epidermis is caught in the drying perspiration and sebaceous secretion, and so remains, as it were, glued to the surface. These different impurities tend to choke the mouths of the perspiratory and sebaceous glands, and prevent their free action. The surface of the body must be kept free from such accumulations.

Much obscurity has been unnecessarily thrown around the subject of the ordinary bath. The rules regarding it are in reality few and simple. The discussions mainly relate to the time, duration, and temperature of the bath, and the results may be summed up thus:

344. Effect of a Cold Bath.—The first effect of a cold bath is to produce a shock to the nervous system, resulting in the contraction of the blood-vessels at the surface of the body. This shock in a healthy person is soon followed by a reaction, in which the heart acts with more vigor than usual, and the contracted blood-vessels are again filled with blood.

The surface of the body then becomes ruddy or rosy, and a pleasurable glow is felt by the bather. The bath should end while this glow of reaction continues, for, if it is allowed to pass away, it is succeeded by a feeling of lassitude and depression, which may last for the rest of the day, and indicates a certain degree of exhaustion of the nervous system. The after-glow may be increased, and the good effect of the cold bath enhanced, by a brisk rubbing of the surface with the hands or a towel.

345. Effect of a Warm Bath.—The first effect of a warm bath is to dilate the superficial blood-vessels and cause a flow of blood to the skin. This produces a general relaxation of the pores of the glands, tends to increase their activity, and in weakly persons there is slight stimulation of the nervous system. After such a bath, the skin being full of blood and in a relaxed condition, exposure to the cold air is dangerous, and the bather should not go out-of-doors until the surface of the body is perfectly dry, and any feeling of languor has disappeared.

346. Rules for Bathing.—Briefly stated, then, these are the rules for guidance in the bath :

Do not bathe within three hours after a meal, as the change in the circulation produced by the change in temperature of the surface of the body interferes with the proper distribution of blood in the digestive organs.

Do not bathe in cold water, if you have found by previous trials that you always have a chilly feeling afterward. This is a matter to be determined wholly by personal experience, and, when any person finds that a particular kind of bath makes him uncomfortable afterward, it is the part of prudence to shun it thenceforth. As a general rule, it may be stated that a vigorous person will feel better after a cold bath, and a feeble person after a warm one, if not too prolonged.

Do not remain in the bath too long. If you feel chilly when you come out, the bath will do you more harm than good. As a general rule, from ten to fifteen minutes is long enough for a warm bath, and five minutes for a cold one.

347. Washing the Hands and Face.—The hands and face are washed more frequently than any other part of the body, because they are more exposed and become dirty sooner. If they can be kept clean with water alone, soap should be avoided, because it unites with the fatty matters on the surface, a certain amount of which is natural and necessary to keep the skin soft and pliable. Soaps that contain an excess of alkali * are particularly injurious, because they not only remove too much of the sebaceous secretion, but irritate the surface and even produce eruptions. After the hands have been washed, especially in cold weather, they should be carefully and thoroughly dried before exposing them to the cold air. It is the neglect of this precaution, together with the excessive use of soap, that causes " chapping " of the hands. The soap removes the fatty matter from the epidermis, thus depriving it of its pliability, and the water, espe-

* Soap is made by boiling fat with potash or soda ; the former makes soft soap and the latter hard soap, such as is commonly used in the toilet. If the proportions of these ingredients are not precisely right, there will be an excess of one of them in the product, and this excess is always alkaline, because special care is taken to neutralize all the fat.

cially if it is warm, is absorbed to a certain extent by the outer cells of the epidermis, so that they swell a little and become softer than before. The epidermis thus becomes less elastic and less tenacious. If it is suddenly exposed to the cold while in this condition, the outer layer contracts and tears apart, in some places making cracks, which often extend down to the derma, and give rise to considerable suffering. Then fatty matter (grease, vaseline, etc.) has to be supplied to take the place of the natural secretion that has been so unthinkingly removed.

To keep the skin of the face and hands smooth and pliable, therefore, soap should be used sparingly,* and they should be thoroughly dried after washing.

348. Care of the Scalp.—The *scalp* should also be kept clean. The epidermis is shed constantly in this region, as well as in other parts of the body, and, together with the dried sebaceous secretion, causes what is known as "dandruff."† The hair should be dressed with a brush that is not stiff enough to scratch the scalp, and the comb should be used only to part the hair and disentangle it, never to relieve itching. Constant scratching of the scalp increases the trouble it is designed to relieve, and makes it rough and inflamed, just as the same treatment will affect the back of the hand or the cheeks. It is well to wash the scalp thoroughly with cold water at least once a week, drying the

---

* The highly-scented toilet-soaps are often made of rancid fat, and injure the skin. The best soap for ordinary use is the white Castile soap, of Spanish or Italian manufacture, which is made of sweet olive-oil and not perfumed.

† Extreme degrees of dandruff are the result of a disease of the scalp, and are often difficult to cure.

hair before going out-of-doors. The hair of girls should never be cut, if it is possible to avoid it, as it is the universal testimony that it never grows as long after cutting as if it is left alone.

349. Care of the Nails.—The finger-nails should be carefully trimmed,* but never cut close to the flesh. If they are cut too short, not only is the delicacy of touch affected, but the ends of the fingers will become club-shaped and ugly. The flesh about the roots and sides of the nails should not be allowed to adhere to them, as it is pulled along and stretched as the nail grows, and produces " hang-nails." The white spots which often appear on the nails are caused by slight blows on the root of the nail during its formation, and therefore are most common in children and in persons whose occupation exposes their hands to such violence. They are of no special importance.

350. Clothing.—The object of clothing is to maintain an equable temperature at the surface of the body. For this purpose the best material to be worn next the skin is a poor conductor of heat. Woolen material, from its porous texture, conducts heat very poorly, and flannel should therefore always be worn next the body, summer and winter. No other material can be compared with it for protecting the surface against sudden changes of temperature. Linen, on the other hand, is such a good conductor of heat, that its use next the skin is absolutely dangerous. Cotton is intermediate between the two. As for the outside clothing, that

---

* It is better to soften the nails slightly in warm water before cutting ; otherwise they will often split or break at the edge, especially in cold weather.

may be safely left to be decided by the comfort and taste of every individual.

The clothing that has been worn during the day should be entirely removed at night on going to bed. The under-clothing, which has been for so many hours in contact with the surface of ti.e skin, is more or less loaded with the matters cast off by it, and, if it is not changed daily, should at least be thoroughly aired during the night.

A shoe that can not be worn continuously during the waking hours from the time it is first put on, should not be worn at all. The habit of "breaking-in" new shoes inevitably results in distortion of the feet and the development of corns. High heels are also bad, because the weight of the body drives the toes down into the narrow end of the shoe, and causes the painful and unnecessary affection known as "in-growing" nails.

Always wear, therefore, a good non-conductor of heat next the skin.

Never wear any article of dress that pinches, for it will inevitably result in distortion or disease, or both, especially in the young, whose bones are yielding and whose organs are not fully formed, and need perfect freedom for their healthy development.

PART VI.

ORGANS OF PERCEPTION.

CHAPTER I.

TOUCH—TASTE—SMELL.

351. Structure of the Papillæ. — The sense of *touch* has its seat in the papillæ of the skin. The nerves, on entering the papillæ, end in certain oval-shaped bulbs, which are very small, and are composed of connective tissue, quite firm in texture, but containing a soft material at the center. The nerve, on entering this bulb, loses its medullary substance (or myelin), and only the axis-cylinder is continued into the center. The papillæ do not all contain such bulbs or corpuscles, but, as they are found in the greatest numbers in places where the sensibility is most acute, it is reasonable to suppose that their function is intimately connected with the sense of touch.

352. The Sense of Touch.—By this sense we gain our knowledge of hardness and softness, heat and cold, roughness and smoothness. The sensitiveness of the skin, as every one knows, varies very much in different parts of the body, and attempts have been made to invent some means of measuring accurately the perfection of this sense. The most

usual method is to apply the points of a pair of
compasses to various spots on the surface of the
body, and see at what distance apart the two points
can be perfectly perceived to be two and not one.
At the tips of the fingers they can be separately
distinguished when only one eighteenth of an inch
apart, while at the small of the back they must be
separated to a distance of nearly two inches before
they will cease to seem one point. Between these
extremes there are many intermediate grades of
sensation.*

353. Deception by the Senses.—It is a remarkable fact that the extremes of any kind of sensation
can not readily be distinguished from each other.
A piece of iron at a white-heat and a frozen solution at 70° below zero will produce much the same
sensation in the part touching them. But it is not
so familiar a fact that, after all, the conscious sense
really lies, not in the termination of the nerve, but
in the nerve-center which receives the impulse from
it, and that, if this be out of order, the sense may be
deceptive. This fact of our liability to be deceived
by our senses applies to all of them—the sense of
touch as well as to others. A single illustration
will suffice. It might seem as if any person would
be able to determine by his feelings whether he
were hot or cold, and yet it is well known to physicians that, in the chill of fever and ague, when the
sufferer, with blue lips and chattering teeth, wants

---

* The most delicately sensitive part is the tip of the tongue, where
the two points are distinguished when only one thirtieth of an inch
apart. As every one knows, a fine hair, which can not be felt by the
finger, or only with the greatest difficulty and uncertainty, is clearly
perceived by the tongue.

to be covered deep with blankets, his temperature by the thermometer is much higher than in health, often running up to 106° Fahr.*

354. The Special Senses.—The *special senses* are all modifications of the sense of touch, and were probably developed out of it. They differ, however, from it, and from one another, in the fact that in each a special kind of sensation is roused in response to whatever form of stimulus. Any stimulus to the retina of the eye, for example, produces only the sensation of light—no matter whether it be light, or a blow, or an electric current—and so with the other special senses (excepting that of taste, as hereafter mentioned). In the eye and the ear, the apparatus for the special appreciation of stimuli forms a highly complicated and delicate mechanism. In the description of these special senses, we shall begin with the one which is least removed from the sense of touch, and end with the one in which sense-perception is the most highly developed.

355. The Tongue.—The sense of *taste* differs less from ordinary sensibility than any of the other senses. It is located chiefly in the tongue, but it has been found, by careful experiments, that we can perceive the flavor of substances also with the soft palate and a small portion of the throat. The *tongue* is a muscular organ of great mobility, and is supplied with three nerves from the brain. One of

* An easy way to satisfy one's self of the deceptive nature of our sensations, and of the fact that they must often be corrected by the judgment, is this : Put one hand into cold water, and the other into warm ; take them out quickly, and plunge them both into water a little cooler than the warm water just used ; it will be found that this water will seem cold to one hand and warm to the other.

these nerves presides over its motions, while the
other two are sensitive nerves, one of which is distributed to the anterior two thirds of the surface of
the organ, and the other to the posterior portion
(Fig. 63). The exact method in which these nerves
terminate is not known, but the surface of the tongue

FIG. 63.—Diagram showing the distribution of the nerves of taste in the
tongue.

is covered with small papillæ, which can be easily
seen with the naked eye, and the delicate extremities of the nerves are known to terminate in these
papillæ. It is well known that no substance can be
perceived by the taste unless it is soluble, and this
renders it almost certain that the matter tasted
comes into actual contact with the end of the
nerve.

356. The Sense of Taste.—The nerves of taste,
in distinction from most nerves of special sense,
also convey sensation like the nerves of general
sensibility. This fact makes it a little difficult to
distinguish between things *felt* by the tongue and
things *tasted*. It is considered by many that there
are in all only *four modifications of the sense of taste*,
viz., *sweetness, sourness, saltness*, and *bitterness*. To

19

these qualities, or savors, others add such savors
as *alkaline, styptic,* etc., which are compound sensations, and it is, to say the least, very doubtful
whether they should be classed as distinct savors.
Qualities which are really the result of feeling, and
not of taste, although perceived by the tongue and
palate, are *pungency, starchiness, piquancy, oiliness,* etc.
Another source of error in estimating savors lies in
the fact that there is a communication between the
nose and mouth through the throat. It is sometimes very hard to distinguish between what we
perceive by the sense of smell and what we taste.
This is in some degree the case with liquids which
have an *aroma* or *bouquet,* but it may be shown that
the *aroma* is perceived entirely by the sense of smell,
by closing the nostrils while tasting the aromatic
substance. It will then be found that the aroma
disappears, although the communication between
the nose and throat still remains open. The reason
for this will be explained under the section describing the sense of smell.

357. Peculiarities of the Sense of Taste.—It has
been stated that the front and back part of the
tongue are supplied by different nerves. This state
of things gives rise to a curious phenomenon, viz.,
that certain substances have a decidedly different
taste when placed on one part of the tongue from
what they have on the other. A few examples of
this will suffice :

If *potassium chloride* be put upon the *anterior* portion of the tongue, it tastes *saltish,* but on the *posterior, sweetish. Sodium sulphate,* on the *anterior* portion,
is *salt,* on the *posterior, bitter. Alum,* on the *anterior*
surface, tastes *acid* and *styptic,* while on the *posterior,*

it has a decidedly *sweetish* taste, with no acid quality whatever.

The sense of taste persists for a short time after the substance tasted is removed. This is probably because the portion of the substance which penetrated the mucous membrane and affected the nerve has not all been taken up and carried away by the circulation. For this reason, it is impossible to appreciate the true savor of different substances by tasting them in rapid succession. The impression made by the first should be allowed to disappear, and only remain in the memory, before the second is tasted. This only applies to delicate flavors. Where the flavors are coarse and pronounced, the slight masking of one by the other often interferes very little with a correct judgment of their character.

The flavor of any substance is perceived in exact proportion to the amount of surface affected by it. So we frequently see persons smacking their lips and pressing their tongues against the roof of the mouth, the better to appreciate a flavor, by spreading the morsel over a larger surface. The use of the sense of taste in the detection of impure and unwholesome food has already been referred to. In addition to this function, it affords us great physical pleasure, although it is generally thought to be of a low order.

The sense of taste, like that of touch, requires actual contact with the substances tasted, and can give us no knowledge respecting substances at a distance from our bodies. We now come to a sense which is a step higher in the scale, for it gives us knowledge of substances at a distance, and it does this by means of material emanations which often can not be detected in any other way.

358. The Sense of Smell.—The sense of *smell* resides in the mucous membrane lining the upper portion of the cavity of the nose. This portion of membrane is supplied by the *olfactory* nerve, which does not convey ordinary sensation, as do the nerves of taste, but carries to the brain only the special sensation of smell. The nerve is distributed to the upper third of the nasal passages, and the rest of the interior of the nose is supplied by nerves of ordinary sensation (Fig. 64).

The amount of any substance required to affect the sense of smell is inconceivably minute. There

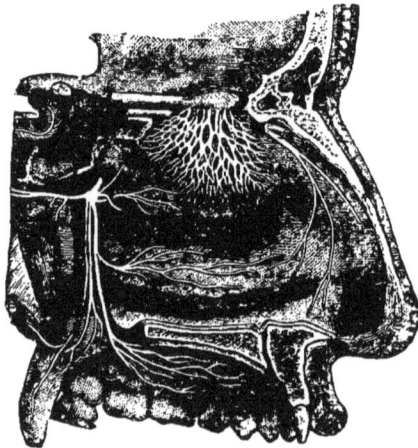

FIG. 64.—The interior of the left nasal passage. The fan-like expansion of nervous fibers is the olfactory nerve. The other nerves shown are nerves of ordinary sensation.

is no chemical or physical means known by which it can be detected in many cases. A grain of musk will perfume a room for months, and lose nothing apparently in weight.

In order to affect the sense of smell, the odorous air must be drawn through the nose. If the air in the nasal passages remains stationary, we smell nothing. This is the reason why closing the nostrils prevents our appreciating aromatic substances in the mouth. For this reason, also, we sniff at whatever has a delicate or faint odor, and, by increasing the rapidity of the air-current through the nose, we add to our appreciation of the odor.* This is very noticeable in the lower animals.

It is necessary to distinguish between substances which really affect the sense of smell and those which merely irritate the mucous membrane. Many substances affect both senses—that of smell and that of general sensibility. *Pepper*, for instance, has a clearly distinguishable and peculiar odor of its own, which is appreciated by the olfactory nerve, and it also irritates the nerves of general sensibility. *Ammonia*, if pure, is merely an irritant, and has no proper odor of its own.

359. Relative Acuteness of the Sense of Smell.—The sense of smell among civilized persons is not educated, and is rather defective. Among savages it is much more acute. Humboldt states that certain South American Indians can detect the approach of a stranger, in a dark night, by the sense of smell, and will also tell whether he is a white man, an Indian, or a negro. But the lower animals far surpass man in this as well as in most other senses. The keenness of the dog's scent is pro-

---

* The distribution of the olfactory nerve in the upper part of the nose also makes it necessary to sniff at faint odors, for otherwise the air containing them might pass through the lower passages of the nose into the throat, without reaching the nerve of smell.

verbial, and he depends much more upon his sense of smell, for recognition of his master, than on his sense of sight.

The sense of smell, like the sense of taste, acts as a sentinel to guard against the introduction of improper food into our stomachs. It also warns us of impurities in the air. After long exposure, however, to a particular odor, the sense appears to become blunted, and that even when the odor is a peculiarly disagreeable one. It seems as if, when its warnings came to be disregarded, it ceased to give any more.

The sense of smell appreciates a great number of distinct odors, and from delicate and fragrant ones we receive a great deal of pleasure. This sense, undoubtedly, occupies a higher plane than the sense of taste, and a delight in pleasant perfumes marks a higher sense of æsthetic gratification than a devotion to the pleasures of the table.

360. The Sense of Hearing.—As we ascend in the scale, we come to a sense which does not inform us of what is taking place in the world outside of us, by actual contact with matter, like the senses of touch and taste; nor by contact with emanations from matter, like the sense of smell; but impressions are produced upon it by means of motion in the atmosphere which surrounds us. This is the sense of *hearing*. Every movement of every form of matter gives rise to waves or vibrations in the air, which lies in contact with the moving substance. These atmospheric waves are received, and the impression made by them is transmitted to the brain by a special apparatus.

361. The External Ear.—The ear (Fig. 65) is usually described by anatomists as consisting of *three* divisions—the *external* ear, the *middle* ear, and the *internal* ear. The appearance of the *external* ear is familiar to all. It has a somewhat shell-like form, has numerous ridges and hollows, and is composed principally of cartilage covered with skin. This external ear serves to collect the waves of sound, and direct them toward the internal parts. In animals this organ is very movable, and, during their waking hours, is generally in motion, to catch sound from various directions. In the human be-

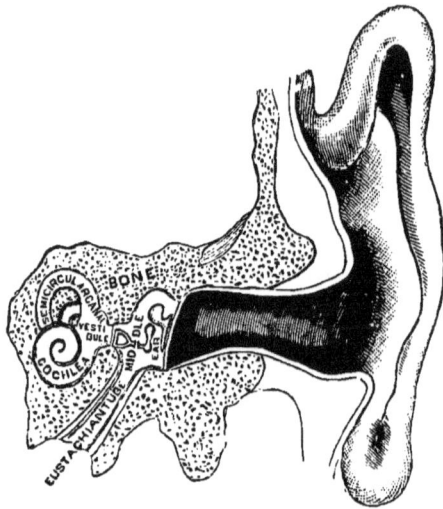

FIG. 65.—Section of the ear, showing the relative positions of the external, middle, and internal ear.

ing there are three small muscles attached to the external ear—one behind, one above, and one in front. These muscles, however, are practically useless, and very few persons have any voluntary control over them. Those who do, can cause the muscle in the rear to contract, and move the ear slightly backward, but the motion is limited, and answers no purpose.

362. The Middle Ear.—From the external ear a canal passes directly inward, toward the interior of the skull, for a distance of about an inch and a quarter. At its inner extremity it is closed by a thin membrane, called the *membrane of the tympanum*, or the *drum* of the ear; and on the other side of this membrane is a small cavity in the bone, about a third of an inch long, a quarter of an inch in height, and a sixth of an inch wide, called the *tympanum*,* or *middle ear.* The membrane of the tympanum is stretched like the head of a drum, and answers somewhat the same purpose. Upon this membrane the waves of air strike, causing it to vibrate ; and this vibration of the membrane is the first step toward bringing the air-waves, or the impulse given by them, into contact with the nerve of hearing. And here we remark that, in order for the vibration of a membrane to be perfect, the air on both sides of it must be of the same degree of density during the vibration. In the common drum this is provided for by having a hole in the side, to allow free passage for the air in and out. In the cavity of the middle ear it is provided for by a short tube, called the *Eustachian†* *tube,* leading into the throat. This tube is about an inch and a half or two inches long, and, for perfect hearing, it is necessary that the air should pass in and out through it with freedom.

363. The Bones of the Ear.—In the cavity of the middle ear is a *chain of minute bones, three* in number, which altogether only weigh a few grains. One of them is attached to the membrane of the tympanum, another is attached to another membrane stretched in drum-like form across a small hole on the opposite side of the middle ear, and the third unites the other two. These small bones have very minute muscles attached to them in such a way that by their contraction or relaxation the bones are made to assume slightly different positions, and the membranes

* *Tym'panum*, a Latin word meaning *drum.*

† *Eusta'chian*, from Bartolomeo Eustachi, a celebrated Italian anatomist (died 1574), who first described it.

to which they are attached are made more or less
tense, and therefore more or less sensitive to sounds.
It is probable that these muscles are somewhat under the control of the will, and that some of the
effort of which we are conscious when we strain our
attention to detect a faint sound, is due to their
active contraction.

364. The Internal Ear.—The second membrane
above spoken of, which is smaller than the membrane of the tympanum, closes the passage between
the middle and internal ears. The *internal* ear or
labyrinth is exceedingly complicated, and the functions of its parts are not yet well understood. The
difficulties of investigation are immense, partly on
account of the minuteness of the organs and partly
because they are situated so deeply in the bones of
the skull and are so near the brain. It is sufficient
to say that the internal ear is made up of many
winding channels and spiral tubes, which go by the
names of the *cochlea*, the *semicircular canals*, and a
cavity called the *vestibule*, with which the others
communicate. All these passages are filled with a
watery fluid, and on their walls and through their
interiors are distributed the filaments * of the nerve
of hearing, the *auditory nerve.* As has been said, the
cavity of the vestibule is separated from that of the
middle ear by a thin membrane, to which one of
the bones of the middle ear is attached. Thus, any
vibration or impulse imparted to this membrane
produces a corresponding pressure on the nerve of
hearing, through the medium of the watery fluid
which fills all parts of the internal ear.

365. Recapitulation.—Thus the air-waves enter-

---

* *Filaments*, small, delicate, thread-like fibers.

ing the external ear strike upon the membrane of the tympanum and put it in vibration. These vibrations are communicated through the chain of bones to the inner membrane, which is also made to vibrate. These vibrations, again, are communicated to the watery contents of the internal ear, and they in turn press directly upon the extremities of the auditory nerve, which conveys the impulse to the brain. And in this way we hear sounds.

366. Different Qualities of Sound.—When we hear sounds, we distinguish many different qualities in them, such as *pitch, quality, timbre, degree,* etc. These differences mainly depend upon the different length, rapidity, height, and character of the air-waves. They are perceived, in a greater or less degree, by every person, but the ease and accuracy with which they are appreciated may be increased by education. The appreciation of musical sounds, for instance, may be immensely increased by practice and proper training.

367. Determination of the Source of Sound.— Besides the perception of the foregoing differences in sounds, which is partly the result of natural endowment and partly that of training, there is another fact with regard to sound which has given rise to considerable discussion. It is the power we have of determining with more or less accuracy the locality from which a sound comes. Some physiologists have supposed that, as the semicircular canals, three in number, are always placed at different angles, they serve in some way, not easily explained, to indicate to us the direction from which the air-waves come.* This view has been supported by

* It has been lately suggested that the semicircular canals have

arguments of some plausibility, but it is now the accepted view that our knowledge of the direction and distance of the source of any sound is the result of past experience. We judge partly by the loudness of a familiar noise, partly by the greater impression on one ear than on the other, partly by the difference in the sound when we turn the head a little one way or the other, the direction of the wind, and a hundred other influences, which have become a part of our experience. If we can not, for any reason, use enough of these methods, we find ourselves unable to locate even a loud and distinct sound; e. g., it is very difficult to tell in what part of a closed room to look for a chirping cricket.

368. Ventriloquism.—The art of *ventriloquism* depends upon an adroit use of these methods in such a way as to deceive. The ventriloquist slyly robs us of all the means which generally serve us to detect the origin of a sound. He speaks without moving his lips; he modulates his voice so that it appears to come from a distance, on account of its faintness; he calls the attention of the spectator, either by word or by gesture, to the point from which he wishes him to expect the sound, and makes clever use of the slight shades in timbre and pitch, which all of us can distinguish, but which it requires great skill and practice to reproduce as

something to do with the sense of direction, and with the preservation of the bodily equilibrium. It has been found that certain obscure nervous symptoms, attended by inability to maintain equilibrium, have resulted from injury to these parts. There is a disease of these organs, called Ménière's disease, from the man who first described it, in which giddiness, as well as deafness, is a prominent symptom. It is probable that the frequent attacks of vertigo, with increasing deafness, from which Dean Swift suffered, were caused by this disease.

professors of this art are able to. The common idea that a man's voice can actually be projected or thrown into a spot twenty feet from his larynx is the idea of pure ignorance, and, as soon as one understands the mechanism of the voice, is seen to be as impossible as for any one to see through a solid stone wall, a thick envelope, a book-cover, or anything else through which no light can pass.

369. Care of the Ear.—The ear is, in almost all its parts, a very delicate organ, but, excepting the external ear, it is so deeply set in bone that it is not very liable to injury; and yet neglect of the organ or ignorant tampering with it may result in permanent and irreparable harm. The tube leading inward from the external ear, called the *external auditory canal*, is sometimes entered by insects. Its walls are moistened by a peculiar secretion, called "wax," produced by small glands just beneath the skin which lines it. This wax is somewhat sticky, and is intensely bitter, and, together with certain short hairs near the outlet of the canal, serves to protect the membrane of the tympanum from the inroads of insects.* The wax has sometimes a tendency to accumulate and interfere with the hearing; and some persons are in the habit of cleaning their ears with ear-scoops. This is an exceedingly dangerous practice, and can not be too severely condemned. The ear is a delicate or-gan, and must be treated delicately by a person who thoroughly understands its anatomy.

370. Danger of Colds in the Head.—The dan-

* Insects can sometimes be coaxed out of the ear by holding a light in front of, and quite near, the external opening. They are attracted by the light, as moths are, and turn round and crawl toward it.

ger of injury to the hearing apparatus through the Eustachian tube is principally from inflammation. Colds in the head almost always affect the hearing in some degree, on account of the swelling of the membrane around the mouth of the tube. But the inflammation may not stop at this point. It sometimes travels along the tube to the middle ear, and when the exceedingly delicate mucous membrane of the middle ear becomes affected, there is usually more or less permanent impairment of the sense of hearing. The little chain of bones becomes stiffened by the disease, just as an elbow or knee does after an attack of rheumatism, and so they conduct the vibrations of the membrane of the tympanum with less force and accuracy. On the slightest indication, therefore, of inflammation of the Eustachian tube, as indicated by unusual dullness of hearing, during an acute catarrh of the nose or throat, a physician should be consulted. The process just described gives rise, in the climate of New York, to about two thirds of all the cases of deafness.

Owing to the congestion of the mucous membrane of the nose and throat produced by alcohol and tobacco, it is found that deafness of varying degree is quite common among drinkers and smokers.

371. The Sense of Sight.—We now come to the highest and most perfect of all the senses, that of *sight*. The four senses previously considered all depend on material contact, in a greater or less degree, for their appreciation of the external world. Even the last sense considered, that of hearing, depends on contact with waves in the air, which is a form of matter, although attenuated. But we now have to do with an organ which perceives external objects through the intervention of light, and light is something which can not be felt or weighed, or detected in any other way than by sight. There have been many theories regarding the nature of light, but the one most in favor at the present day is the *wave theory*, or *undulatory theory*, so called.

372. The Nature of Light.—According to this theory, the universe is pervaded by an exceedingly subtile form of matter called the *ether*, and light consists of waves propagated through this ether with tremendous rapidity. The average number of vibrations in a second is estimated to be over 500,000,000,000,000, and they travel at the rate of about 187,000 miles per second. The wave theory of light is only a theory, and the existence of the extremely attenuated form of matter called ether, which can neither be weighed, seen, heard, nor felt, but only postulated from certain phenomena supposed to depend upon it, is only inferred; but still this theory explains almost all the phenomena of light, and, until some better one is suggested, must be retained as a working hypothesis.

From these facts, we can judge of the extraordinary delicacy of the eye as an organ of sense. Here is a stimulus, of unknown origin, which can traverse the widest regions of space, which enables us to appreciate the existence and form and even the structure of bodies which are billions of miles away from us, and gives us more knowledge of the external world, perhaps, than we obtain through all our other senses together, and yet its real nature is as hidden and mysterious and impalpable as the nature of our consciousness. Let us now examine the structure of the organ which perceives this strange stimulus and enables it to affect our brain.

373. Situation of the Eye.—The *eye* is situated in a cavity called the *orbit*, surrounded by bone excepting in front, and padded all about with fatty and muscular tissues, so that, although it is protected from injury by an unyielding bony case, it still reposes on a soft and elastic bed.

374. Structure of the Eyeball. — The *eyeball* (Fig. 66) is nearly spherical, and about an inch in diameter. It is formed of three membranes, arranged concentrically one within the other, and the interior is filled with certain structures necessary to vision. The outer membrane is called the *sclerotic**

---

* *Sclerot'ic,* from the Greek σκληρός, *hard,* because it is the hardest and toughest coat of the eye.

coat, and in front, where it is visible, constitutes what is commonly known as the white of the eye.

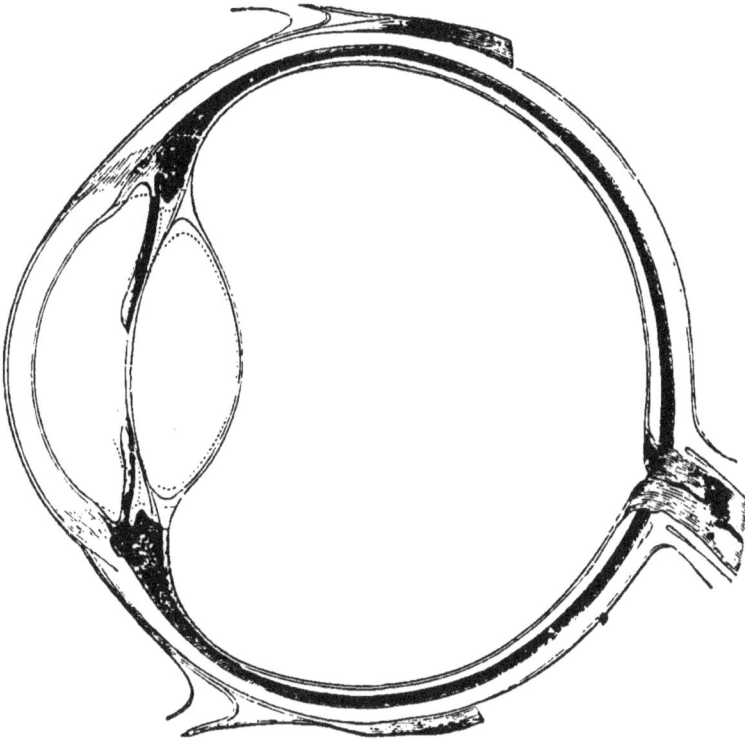

FIG. 66.—Horizontal section of the right eye, showing the relative position of its parts.

It is composed of white, fibrous tissue, and is exceedingly strong, tough, and elastic. It surrounds the whole eyeball, excepting in front, where a transparent membrane, about $\frac{1}{25}$th of an inch thick, is set into it like a watch-glass in the case. The front of the eye must of course be transparent for the admission of light. This portion of the eye is called the *cornea.*[*]

[*] *Cor'nea*, from the Latin *cor'nu, a horn*, because of its resemblance to transparent horn.

Just inside the sclerotic coat is another coat, called the *choroid*,* which covers the whole interior of the eye, excepting that portion bounded by the cornea. This coat is very plentifully supplied with blood-vessels, and with immense numbers of minute cells, filled with coloring-matter, so that it appears of a dark-brown or chocolate color, and in some persons almost black. Inside this choroid coat, again, is the *retina*,† which is an exceedingly delicate and complicated membrane made up of nervous tissue, and upon which the impressions of light are received. The retina is spread over about two thirds of the inner surface of the eyeball, not reaching quite as far forward as the sclerotic and choroid coats.

375. The Internal Parts of the Eye.—The membranes above mentioned form, so to speak, the incasement or shell of the eyeball. The bulk of the interior of the eye is formed by the *vitreous humor*.‡ This is a soft, semi-fluid, transparent, jelly-like body,#

---

* *Cho'roïd*, from the Greek χόριον, *leather*, because, being dark-colored, it resembles leather in appearance.

† *Ret'ina*, a Latin word, meaning in that language just what it means in English. It is derived from *re'te, a net*, on account of its mesh-like appearance.

‡ *Vit'reous*, from the Latin *vit'reus, glassy*.

The vitreous humor contains, even in healthy eyes, minute bodies, which can not be detected from the outside, but which can be seen, greatly magnified, by the person himself. They are called "*muscæ volitantes*" (literally, *flitting flies*), and look like small strings of bright beads, or little transparent spheres or fibers, which move when the eye is moved, and, when the eye is held perfectly still, seem to sink slowly. They really float up toward the top, but appear to go down, their direction being reversed by the optical apparatus of the eye. These little objects are seen most vividly against a bright surface, like the sky, or a white wall, and when perceived for the first time are apt to frighten

which fills the ball of the eye, with the exception
of a small part, about one sixth of the mass, in front,
which contains the *crystalline lens* and the parts between the lens and the cornea. This vitreous humor is surrounded by a very delicate membrane,
also transparent, which lies in immediate contact
with the whole extent of the retina. Just where
this membrane passes in front of the vitreous humor, and just behind the cornea, it splits in two
layers, and between these layers is suspended and
held in place the *crystalline lens.* This lens is a
double-convex one, with the posterior curvature a
little greater than the anterior, is about one third of
an inch in diameter from side to side and a quarter
of an inch thick at its middle. It is, of course, perfectly transparent, and acts precisely as a double-
convex lens acts in any optical instrument. Just in
front of the lens is a curtain, with a hole in its center, which serves to regulate the admission of light
to the interior of the eye. This curtain is called
the *i'ris,** and it contains muscular fibers, to which
it owes its power of contraction, and pigment-cells,
to which it owes its color, varying in different persons. The hollow space between the cornea in
front and the crystalline lens behind is filled by a
fluid called the *aqueous humor,†* composed almost
entirely of water, with a little salt.

376. Uses of the Two Outer Coats of the Eye.
—The *sclerotic* coat serves, by its toughness and

people. They exist, however, in every eye, and are perfectly harmless.

*'Ιρις, a Greek word, meaning *the rainbow*, so called on account of
the variety of colors it presents in different eyes.

† *A'queous*, from the Latin *a'queus, watery*, because it consists almost entirely of water.

elasticity, to give shape to the organ, and protect the parts within. The *cornea* in front answers the same purpose, and is transparent, in order to allow the passage of light. The *choroid* coat serves as a nest for the blood-vessels which nourish the retina, and also, by its dark color, prevents the rays of light, which have once passed through the retina, from passing back again, and so confusing the sight. When the light which enters the eye is so intense that it can not be absorbed by the choroid, it is thus reflected through the retina, and our sight is not clear, as we have a double impression, coming from two directions, and the rays conflict with each other. We call this " being dazzled." In albinos, the coloring-matter of the choroid is absent, and such persons always are troubled with dimness and confusion of sight.*

FIG. 67.—Vertical section of the retina, highly magnified. At the upper part of the cut are the rods and cones, while below are several distinct layers of nerve-cells.

* Albinos not only can not see well, because they are dazzled by the light, but their eyes have a constant vibratory motion from side to side, like the pendulum of a clock. This peculiar affection is known to physicians as *nystag'mus.*

377. The Retina.—The *retina* is formed by the expansion of the *optic nerve*, which enters the eye behind and spreads out over the interior. It is exceedingly complicated in its structure (Fig. 67), no less than eight distinct layers being found in its thickness, although the whole taken together is very thin and delicate. The outermost layer is made of innumerable minute cylinders of nervous matter of different shapes and sizes, packed together side by side like the seeds of a sunflower. These are called the *rods* and *cones* of the retina. Inside of these are other layers of tubes, fibers, cells, and granular matter, all of which, doubtless, have their part to play, but the particular function of which is not yet known. This arrangement of the nerve substance is necessary to the sense of sight, for, singularly enough, the spot where the optic nerve enters the eyeball is entirely blind. This nerve can convey the impression of sight to the brain as it receives it from the special sense-organ the retina ; but, if light falls directly upon the optic nerve itself, no sensation is produced. This curious and interesting fact can be easily demonstrated.

Make a round black spot and a black cross upon a white card, three inches apart (Fig. 68). Now

FIG. 68.—Diagram to demonstrate the existence of the blind spot.

hold the card in front, at a distance of about a foot from the eyes, close the left eye, and look at the cross with the right one. Both the cross and the black spot will be seen distinctly. Now move the card slowly toward you, still keeping the right eye fixed upon the cross. At a certain point the round spot will disappear, but, as the card continues to approach the eye, it will reappear within the field of vision. At the point of disappearance, its image falls upon the optic nerve just where it enters the eye (Fig. 69). Now, if this blind spot were just in the axis of the eye, we should be badly off, for it is

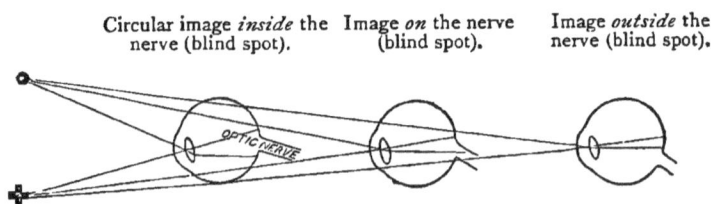

FIG. 69.—Diagram illustrating the blind spot. It represents a horizontal section of the right eye, the axis of the eye in every case being turned toward the cross.

evident that we should be unable to see anything we looked directly at. But the point of entrance of the optic nerve is toward the inner side of the eyeball, and consequently we are not incommoded at all by this small spot, which is insensible to light.

378. Use of the Crystalline Lens.—The interior of the eyeball is, therefore, like the interior of a *camera obscura.* It forms a dark chamber, lined with a dark membrane, to absorb superfluous light, and with a small opening in front to admit light. Just behind this opening is the lens.

If light came directly into the eye, and fell upon the retina, without being brought to a focus, our brains would appreciate the existence of the light, but would not get clear ideas of the appearance of objects. Everything would be dim and confused.

Something of this may be seen in a camera obscura, if the lens be removed. The rays of light, as they enter the camera, still form an image of outside objects, but it is a dull, indistinct, and obscure one. With the lens, however, the picture is bright and distinct in all its details. The reasons for this are too long to be stated here, and will be found in any treatise on optics. Suffice it to say that the function of the *crystalline lens* is to concentrate the rays of light as they enter the eye, and bring them to a focus on the retina.

379. The Function of Accommodation. — The rays of light being affected by this lens precisely as they would be by a glass one, there must be some provision made for the varying distance of objects from the eye. In the camera, as objects are nearer or more distant, we draw out or push in the lens, so as to bring it farther from or nearer to the surface which receives the image. Now, we find that, within certain limits, a healthy eye sees objects a moderate distance away with just as much distinctness as those near at hand. Moreover, we are conscious, as we change rapidly from looking at a distant object to one close by, of a kind of effort in the eye itself. There must be a change of some kind there, corresponding to the pulling out or pushing in of the lens of the camera. What is the change?

There is a short, delicate muscle, called the *ciliary muscle*, one extremity of which is attached to the stout membrane of the sclerotic and cornea at their junction, and the other extremity to the choroid. The muscle is a circular one, reaching all around the eyeball, and is only about an eighth of an inch broad. Now, the crystalline lens is very elastic, and, as has been stated, it lies between two layers of a membrane which passes backward and surrounds the vitreous humor. Under ordinary circumstances, these two layers of membrane, one of which is in front of the lens and the other behind it, are supposed to exercise some pressure on it, and render it a little flatter than it would be, if it were left to take its own shape, in accordance with its elasticity. Now, suppose that we wish to look at an object close at hand. The ciliary muscle contracts and the membranes surrounding the vitreous humor are drawn forward slightly. As a consequence of this the two layers of membrane, between which the lens lies, are somewhat relaxed, and exert less pressure on the front and back of the lens. The pressure being removed from it, the elasticity of the lens makes it assume a more convex shape (Fig. 70), and consequently brings the rays of light to their proper focus. This is supposed to be the method by which the eye accommodates itself to different distances, and its operation is so perfect and exact that, within certain limits, we can see whatever we wish to. Beyond a certain distance, and within a certain distance, this accommodation no longer occurs. These distances vary with different persons, and the inner limit is called the *limit of distinct vision.*

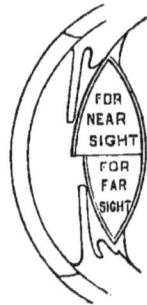

FIG. 70.—Diagram showing how the lens changes its form for near and far sight.

380. The Limit of Distinct Vision.—The limit of distinct vision depends on the accommodation of the lens above described. As any object is brought nearer and nearer to the eye, the effort required to see it distinctly, or, in other words, the effort at accommodation, is greater and greater, until at length it begins to be accompanied by pain and a peculiar sense of fatigue in the eyes. Within this point distinct vision is not possible, for the ciliary muscle can contract no further. In healthy eyes, this limit is usually about six inches from the cornea.

As objects recede, the rays from them come to the eye at a smaller and smaller angle, until, at the distance of fifty feet, they are almost parallel; that is, they are practically so, as far as the perception of the retina is concerned. Beyond this point, vision is distinct enough, and no accommodation of the lens is required; but a new difficulty comes in to hamper us in our perception of objects. We have seen that, however delicate the sense of touch may be, it is possible to press two points of a compass on the skin so near together that we are not able by the sense of touch to say whether they are two points or one. A similar condition is found to exist in the retina. It has been shown by experiment that, if two objects or two points of any object are so near together that both combined subtend an angle of less than one minute at the lens, the retina can not distinguish them apart. Beyond the distance of fifty feet, then, without artificial assistance, the eye is unable to distinguish objects perfectly in their minute details.

381. The Function of the Iris.—The color of the iris is due partly to the blood-vessels which run through it, and partly to small pigment-cells. In new-born children the iris is always blue, and does not take on the color which is to last through life

for several weeks after birth. The iris performs
two functions. In the first place, in every lens, the
whole of which is made of the same substance and
of the same density throughout, the rays of light
which pass through and near the center are not
brought to a focus as soon as those which pass
through near the circumference. This fact causes
a blurring of the image (spherical aberration) and
also a partial decomposition of the light (chromatic
aberration), so that the image appears colored at
its edges. In optical instruments these difficulties
are remedied, partly by constructing the lens of two
different substances, which counteract each other's
defects, and partly by covering the edge of the lens
and only allowing the light to come through the
center and the immediately adjacent parts. The
latter method is the one carried out by the iris. It
is pierced in the center, forming the pupil of the
eye, and this pupil is situated immediately in front
of the center of the lens, so that in this way the
faults above mentioned are corrected. The iris
also regulates the admission of light to the eye.
Too much light irritates the retina, and by reflex
action the iris contracts. If there be too little light
the iris is relaxed, and the pupil becomes larger so
as to admit more. This process has already been
mentioned at length.

**382. The Retina and Optic Nerve sensitive only
to Light.**—The retina is sensitive to other impulses
than that of light, but, whatever these impulses
may be, the sensation of light is the only one conveyed to the brain. If the eyeball be struck or
pressed, or an electrical current be passed through
it, we see sparks or flashes of light, and this even

when we are in a perfectly dark room. This is due to the shock to the retina, from which every stimulus of whatever kind is transmitted by the optic nerve to the brain under the form of light.

383. Persistence of Impressions on the Retina.—When an image is formed on the retina, especially if it is a very bright one, the impression remains a little while after the object that caused it has passed by. The consequence of this is that, if several objects follow each other in very rapid succession, new images are continually formed on the retina before those that immediately preceded them have had time to fade, and so the eye does not detect any interval between them. Thus, when we look at a swiftly-revolving wheel, we do not see the separate spokes, but a continuous, hazy blur; and when a live coal is whirled around fast enough before the eyes, we do not see the coal in its proper shape, but only a luminous circle. In this temporary persistence of impressions on the sense of sight, it is like all the other senses.

384. Color-Blindness.—The power of discriminating between different colors varies very much in different individuals. While some can not only distinguish the colors with ease, but can pick out the most delicate shades of any particular color with unfailing accuracy, others are unable to discriminate, for example, between red and green. Such persons are called color-blind, and can not see any difference between the fruit and the leaves of a cherry-tree, excepting by the shape.

385. Near-Sight and Far-Sight.—In the natural and healthy eye the rays of light are brought to a focus on the retina, but in some eyes they are not. In near-sighted persons the eyeball is usually a little too long, and the rays of light, being brought to a focus before they reach the retina, cross each other, and the image is blurred. In far-sighted eyes the eyeball is too short, or the lens is too flat, and the rays are not brought to a focus at all, so that the effect upon vision is to make everything look blurred, as in the former case (Fig. 71). Both of these defects can and should be corrected by proper glasses, for the constant straining to see, of

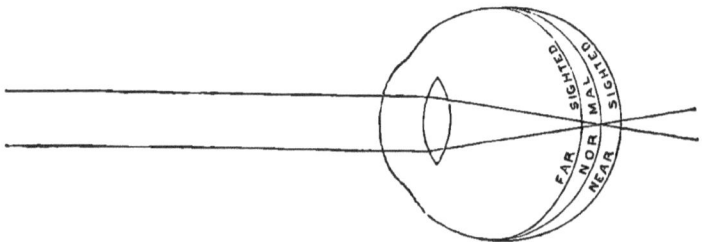

FIG. 71.—Diagram showing the point at which the rays of light are brought to a focus in different eyes.

short-sighted and far-sighted persons who try to get along without glasses, eventually injures the sight.

386. Muscles of the Eyeball.—To the outside of the eyeball are attached six muscles, by means of which it can be moved in any direction. Four of them pass from the four sides of the eyeball straight backward to the bone at the rear of the orbit, and move the eye up or down, and to the right or left. The other muscles are attached, one to the upper and the other to the lower surface of the eye-ball, their other ends being attached to the bones on the inner side of the orbit in such a way that, by their contraction, the eyeball is rolled in one direction or another.

387. The Eyelids and Eyebrows.—This delicate organ—the eye—is protected from injury in front by a number of accessory parts, viz., the *eyelids, eyelashes*, and *eyebrows.*

The *eyelids* are composed of thin pieces of cartilage, covered with skin on the outer and lined with mucous membrane on the inner side. They are movable, and, when open, expose nearly all the iris; when shut, they completely cover the eyeball. At their margins are several short, stiff hairs—the *eyelashes*—which project outward and downward, and serve as an additional protection against the entrance of dust and insects. Above the eyes are ridges of thick skin, also covered with hairs, and called the *eyebrows*, which turn aside the perspiration or other fluids which run down the forehead toward the eyes.

388. The Tears.—In order to keep the cornea perfectly smooth and transparent, and to preserve the softness of the mucous membrane lining the interior of the lids, which continually sweep over the cornea, and if rough would irritate and ruin it, a special fluid is provided, called the *tears*, which are continually secreted, and keep those parts constantly moist. The gland which provides the tears, called the *lach'rymal gland* (Fig. 72), is situated inside the orbit, just above and outside of the eyeball, and its secretion is poured out on the eye by several little openings at the upper part of the upper lid, on its inner surface. By the winking of the lids, which occurs frequently and is usually involuntary, it is spread over the surface of the eye. On the edge of the lids, near the inner angle of the eye, may be seen two minute openings, one in each lid.

These are the mouths of two little canals, which
pass from these points toward the nose, soon unit-

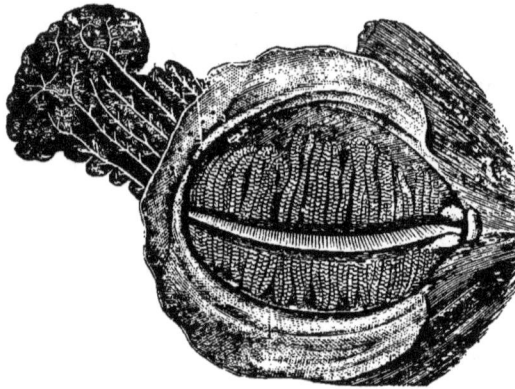

FIG. 72.—Left eyelids from behind. Above and at the left is the lachrymal
gland. On the lids are seen what look like strings of beads. They are
glands that secrete the fatty matter that coats the edges of the lids, and
keeps the tears from running over.

ing in one larger canal, which runs downward into
the nasal cavity. Through these minute canals, the
superfluous tears run off into the nose. These
canals are lined with mucous membrane, and, during a cold in the head, the inflammation often closes
them, and the tears, not finding free passage through
them, overflow upon the cheeks. This overflow
also occurs when from pain or any powerful emotion the tears are increased in amount so that the
canals are unable to dispose of the extra supply.
This phenomenon we call *crying* or *weeping*. An
overflow of tears is guarded against under ordinary
circumstances by an oily secretion of certain glands
in the lids, which is discharged just along their
edges. This secretion keeps the lids from sticking
together when closed, and also retains the tears, unless their quantity is greater than common.

389. Effect of Alcohol and Tobacco on the Eyes.
—When alcohol and tobacco are used habitually, dimness of vision sometimes results, beginning with a slight obscuration, as if there were a fog before the eyes, and gradually progressing, without pain or any signs of inflammation, to total blindness. If the bad habit is abandoned when the first symptoms appear recovery is almost certain. At this period of the disease there is simply some congestion of the optic nerve. But if the warning is unheeded and the drinking or smoking is continued, the nerve begins to waste, and atrophy is the result. This produces incurable blindness. Some careful observers state that this disease, when produced by tobacco, presents peculiar features, which point out its cause, and it has been called "tobacco-amaurosis." This seems to be the only organ of the body in which it is yet positively known that tobacco causes permanent degeneration.

390. Care of the Eyes.—It is exceedingly important that proper care should be taken of the eyes. They are very delicate, and yet there is hardly an organ in the whole body, not even excepting the stomach, that is more frequently abused. If anything be the matter with the eyes—if they smart, or tingle, or itch, or the sight is dim, or blurred, or indistinct—a good oculist should immediately be consulted. More is known about the eye, probably, than about any other organ in the body. The most eminent scientists in the world have spent their best years in the study of it in health and disease, and it can be examined inside and outside thoroughly. As a consequence, operators who devote their special attention to this organ generally know what they are about, and their advice should be carefully and minutely heeded. It will not do for individuals to treat their own eyes, and therefore it is unnecessary to say anything about its disorders. With regard to proper care of the eye, it is almost enough to say, Do not do anything to produce a feeling of "strain" in the organ. This sensation, as already explained, is produced by the effort of accommodation, and, as the muscular contraction accompanying this act compresses the interior of the eye somewhat, it causes a congestion of the blood-vessels, which is usually only temporary, and passes away when the cause is removed. If a strain like this, however, be often repeated or long continued, it is apt to result in permanent injury to the sight, just as long-continued congestion of the blood-vessels in any other part of the body will impair the working of the organ in which it occurs. The same effect is produced by reading or using the eyes for fine work while stooping or lying, or during the incessant jarring of a vehicle, especially a railroad-car. The eyes should not be rubbed, or pressed, or squeezed, or used by a bad light, or in any other way treated as if they could bear rough usage.* They can not, and, if they are abused, the sight is apt to fail in a sudden and alarming manner.

* It is supposed by many that the eyes are strengthened by being opened in cold water when the face is washed. This is not so. On the contrary, it is an exceedingly injurious practice, and frequently produces little ulcers on the surface of the eye, which, although perhaps not dangerous, are very painful, and may prevent the use of the affected eye for several days. The eyeball is washed sufficiently by the tears.

PART VII.

ORGAN OF SPEECH.

391. Structure of the Larynx.—In the upper part of the neck, in front, is a hard, projecting mass, which, in the throats of thin persons, is plainly visible, and in every one can be easily felt. This is commonly called *Adam's apple*, and consists of the projecting cartilages which form the sides of the *larynx*. This organ is situated at the upper end of the trachea, and a cross-section of it is nearly triangular in shape. It is composed mainly of cartilage, and forms a stiff, open, box-like organ, covered on the outside with muscles, and on the inside with mucous membrane.

392. The Vocal Chords.—The upper end of the larynx is nearly closed by muscular and membranous tissues, which divide it at this point from the throat. But it has a chink-like opening, the *glottis*, previously described, situated at the base of the tongue, just behind the *epiglottis*. This is the essential organ of voice. All the air which passes into or out of the lungs must go through the glottis. Now, in front, the *vocal chords*, which form the

21

sides of the glottis, are nearly or quite in contact, while the posterior extremities can be separated to quite a distance from each other. These posterior ends are attached to two small cartilages, which can be rotated by certain muscles, so as to separate the vocal chords. When certain other muscles contract, which act in an opposite direction, the cartilages are rotated inward, and the vocal chords are brought nearer together.* Other muscles render the chords tighter or looser, according to circumstances (Fig. 73).

393. How the Voice is produced.—Now, when

High note.   Medium note.   Low note.

FIG. 73.—The vocal chords in different positions. The opening between the chords is the glottis.

the air is forced through the glottis in expiration with sufficient rapidity, the vocal chords are thrown into vibration, and, vibrations being thus caused in the column of air in contact with them, sound is produced. The quality of this sound, as well as its pitch and tone, depends upon well-known physical laws. The farther apart and the more relaxed the vocal chords are, the lower will be the sound ; the

* These cartilages are triangular in shape, and act very much like the triangular pieces of metal to which bell-wires are fastened. The vocal chord being attached to one corner, and the muscle to another, while it rotates about the third angle, it is evident that when one muscle contracts the vocal chord will be moved outward, and when another opposing muscle contracts the end of the chord will be moved inward.

nearer together and tighter drawn they are, the more acute will be the sound. Differences also depend upon the size of the larynx and trachea, the length of the vocal chords, their thickness and smoothness, the condition of the throat and parts adjacent, etc.

In this way the sound is made which we call *voice*, and it is modulated into articulate speech by the action, separate or combined, of the throat, nose, palate, tongue, teeth, and lips. To consider how these variations are produced would require great space, and be foreign to the purpose of this book. It is sufficient to say that the voice is produced in the larynx, and articulation performed by the parts above that organ.

394. Abuse of the Larynx.—Abuse of the larynx produces bad results, as does abuse of any other organ of the body. Perhaps the most common form of this is the use of the voice, when the vocal chords are in an inflamed or congested condition. Overuse of the larynx in speaking is apt in many persons to bring on congestion of the chords. This causes them to swell a little, and the voice loses some of its clearness. If rest be not given the organ at such a time, the congestion continues, and becomes chronic. Then the voice is permanently altered, becoming rough and harsh as compared with its original quality. Very often a sore-throat extends to the glottis, and the mucous membrane which covers the vocal chords becomes inflamed. Then the usual results of inflammation of a mucous membrane follow. The chords swell and are covered with a thick secretion, like that which is coughed up during such an attack. This thickening of the chords affects the voice, and makes it not only harsh but of a lower note than usual, just as in a violin the large strings give forth the low notes and the smaller ones the high notes. The secretion also hinders the vibration of the chords, and sometimes the voice is reduced to a whisper or even extinguished, because the chords are so disabled as not to be able to vibrate at all. A curious phenomenon sometimes occurs in such cases, called the "breaking of the voice." If the mucous secretion be very viscid, as it often is, the chords may stick together at some point during the act of speaking, and the part that vibrates is instantly reduced in length to perhaps three quarters of what it was a moment before. This has the same effect as the shortening of a string of a violin by pressing it with the finger. It makes the voice suddenly take on a higher note. In the next instant the chords become entirely separated, and the former note returns, and so the speech consists of a singular series of growls and squeaks.

395. Care of the Larynx.—When the chords are in this condition they ought to be used as little as possible, or permanent injury to the voice will be apt to result. It must be remembered that the mucous membrane of the larynx in such cases is diseased and sore, and ought not to be rasped by a forcible current of air, or it will not readily recover. If one has a sore spot on his hand, he does not rub it violently several times a day; he lets it rest until it is well again. In such a case, the pain which results from rough treatment is a sufficient indication of its harmfulness, but inflamed or congested vocal chords are not usually painful, and the use of the voice is so necessary a part of our daily life that every one is too apt to neglect warnings regarding it. But Nature is inexorable, and shows no favor. Her penalties are severe and unerring.

TABLE OF THE BONES.

SKULL :	Cranium (brain-case) :	Occipital (back of head)............ 2 Parietal (sides of head, above)..... 2 Temporal (sides of head, below)... Frontal (forehead)................. Sphenoid (at base of skull) .... Ethmoid (base of skull, over nose)...	8
	Face :	2 Superior maxillary (upper jaw).... Inferior maxillary (lower jaw) ...... 2 Malar (cheek-bones)............. 2 Lachrymal (part of the orbit)...... 2 Nasal (bridge of nose)........... 2 Inferior turbinated (inside of nose). 2 Palatal (at back of nose)......... Vomer (part of partition of nose)....	14

Total for SKULL: **22**

Os HYOIDES (hyoid bone, at base of tongue)........................ **1**

SPINE (back-bone) :	Cervical vertebræ (in the neck) : { Atlas (next to skull). Axis.............. 5 others............ } 7	
	Dorsal vertebræ (supporting the ribs)............. 12	
	Lumbar vertebræ (in small of back, or loins)....... 5	
	Sacrum { (These bones, together with the } 1	
	Coccyx { hip-bones, form the *pelvis*.) } 1	

Total for SPINE: **26**

RIBS (12 on each side) **24**

STERNUM (breast-bone) **1**

UPPER LIMBS :	Clavicle (collar-bone)........................	1	32 in each limb.
	Scapula (shoulder-blade)	1	
	Humerus (upper arm)........................	1	
	Radius } (forearm)........................ Ulna }	2	
	Carpal bones * { Scaphoid, semilunar, cuneiform, (wrist) : { pisiform, trapezium, trapezoid, os magnum, unciform........ }	8	
	Metacarpal bones (hand).....................	5	
	Phalanges (thumb, 2 ; fingers, 3 each)............	14	

Total for UPPER LIMBS: **64**

LOWER LIMBS :	Os innominatum (hip-bone)	1	31 in each limb.
	Femur (thigh-bone).........................	1	
	Tibia (shin-bone) } (leg) .... Fibula.......... }	2	
	Patella (knee-pan)..........................	1	
	Tarsal bones { Calcaneum or os calcis (heel-bone), (ankle) : { astragalus, cuboid, scaphoid, in- ternal, middle, and external cù- neiform................... }	7	
	Metatarsal bones (foot)......................	5	
	Phalanges (great toe, 2 ; other toes, 3 each)........	14	

Total for LOWER LIMBS: **62**

Total †... **200**

* Beginning at thumb-side of wrist in each row.

† The minute bones of the ear, 3 on each side, are sometimes included in the list of bones, making 206. The teeth (20 temporary and 32 permanent) are not bones.

QUESTIONS.

PART I.

CHAPTER I.—1. What is anatomy? Physiology? Hygiene? What is an anatomical element? A tissue? An organ? A system? An apparatus? Give examples of each. What is the function of an organ? What is a poison? A stimulant? A narcotic? A narcotic poison? What is opium? Morphia? Laudanum? Paregoric? How is alcohol produced? Its properties? Its composition? Is it always an artificial product? Tobacco? Its most important ingredient? Properties of nicotine? What is chloroform? Chloral?

CHAPTER II.—2. What is the minute structure of the body? Which is the original element? What causes the different consistency of different organs? 3. What is a fiber? Where found? 4. Why is the cell so important? What is a cell? Nucleus? Nucleolus? 5. What is protoplasm? How large are cells? 6. What other kind of matter exists in the body? 7. What is the difference between living and dead cells? Illustration? 8. How do cells increase in number? 9. What other powers do cells possess? How are wounds healed? What is said of "proud flesh" (foot-note)? Does one cell ever perform the duty of another? Give example (foot-note). 10. How does alcohol affect growing cells?

PART II.

CHAPTER I.—11. Why are bones necessary? 12. How does living bone differ from dead bone? 13. What is the composition of bone? What experiments will show this? 14. How are bones affected by age? What is a "green-stick" fracture? 15. How many kinds of bones are there? What is the structure of the shaft? Of the extremities? Why the difference? What is the marrow? 16. What is the periosteum? What is the minute structure of bone? 17. What is the use of the periosteum? Illustrate.

CHAPTER II.—18. How many bones in the body? What is ossification? When complete? 19. What is the spine? What is a vertebra? How are the vertebræ arranged in the spine? What separates them? How is the spinal canal formed, and what does it contain? 20. How movable is the spine? How is it held together? What is the use of the pads between the vertebræ? Is a man taller in the morning or at night? Why? 21. What is the skull? How are its bones peculiar? What is the advantage of its arched shape? What movable bone in the skull? 22. What are the sutures of the skull? What is their use? 23. What are the frontal sinuses? How may they cause headache? 24. Describe the ribs. How many are there? How are they attached to the breastbone? What are the floating ribs? How does the motion of the ribs alter the form of the chest? 25. How do the cartilages of the ribs change with age? How does pressure affect them? What is the natural shape of the chest? 26. How many bones are there in the limbs? What is meant by homologous bones? 27. How are joints formed? What is cartilage? Synovial membrane? Ligaments? Uses of these parts.

CHAPTER III.—28. What is a fracture? A dislocation? How can they be told apart? 29. How are bones generally broken, and why does the limb generally become shorter after a fracture? Why is a fracture near a joint so serious? 30. Why is a dislocation so painful? In what joints are dislocations most common? How are they generally caused? What is a sprain (foot-note)? 31. How long does it take a broken bone to unite? 32. How should a broken limb be cared for before the doctor comes? How can the patient be carried? 33. How does alcohol affect growth? 34. Effect of tobacco on development? 35. Effect of opium on children?

CHAPTER IV.—36. How are the bones moved? What two kinds of muscle are there? 37. What is a voluntary muscle? What is connective tissue (foot-note)? Use of muscle. How does voluntary muscular fiber look under the microscope? What is striation? 38. What is the minute structure of involuntary muscular fiber? 39. How do voluntary and involuntary muscles differ in their action? Peculiarity of the heart-muscle. Where are involuntary muscles found? 40. Do muscles vary much in size? Illustrate. 41. How are the muscles connected with the bones? What is a tendon? Describe the peculiar arrangement of tendons at the wrist and ankle. 42. Describe the disadvantages under which a muscle exerts its force during contraction. 43. What is the irritability of muscle? Give illustrations. 44. What is meant by the muscular sense? Of what use is this sense? 45. Illustrate the use of this sense in standing. 46. What is the effect of contraction on the muscle itself? 47. What is the effect of muscular overwork? 48. What is the result of muscular inactivity? Give illustrations. How is curvature of the spine produced (foot-note)? 49. What is the best exercise? What is said of gymnastic training? 50. Why is exhaustion dangerous? 51. Why is rest necessary? What simple rules for exercise are given? 52. What permanent changes does alcohol tend to produce in the body? 53. How does it affect the muscles? 54. How does it affect the power of endurance?

PART III.

CHAPTER I.—55. Why do we need food? 56. Can people live without eating? Why? 57. What are the two great divisions of foods? 58. What proportion of the body is water? How much water is needed daily? 59. Illustrate the importance of salt (foot-note). 60. What other inorganic substances are taken in food? Why is lime so important? 61. What are the non-nitrogenous foods? 62. In what foods is starch found? What peculiarity has starch? 63. In what foods is sugar found? Name some of the varieties of sugar. What is said of glucose (foot-note)? 64. What is the use of fat in the body? Is the fat all taken into the body with the food? What articles of food tend to produce fat? Illustrate. Describe Mr. Banting's case (foot-note). 65. What are nitrogeneous foods? What other name are they known by to physiologists? In what foods are the nitrogenous substances found? Are similar substances found in vegetable foods? 66. Why do we need variety in our food? 67. Which is the most essential of all articles of food or drink? Illustrate. 68. How much of each kind of food is needed daily? 69. What is the effect of cooking upon food?

CHAPTER II.—70. What is fermentation? Its cause? 71. Mention some of the different kinds of fermentation. Describe vinous fermentation. What are its chief products? Describe acetous fermentation (foot-note). Is alcohol a food? 72. What are alcoholic beverages made of? 73. How are malt liquors manufactured? 74. What are the principal malt liquors, and how do they differ? 75. What peculiar and dangerous quality has alcohol, in common with other narcotic poisons? What is the only absolute safeguard against it? 76. How are domestic beers made? Are they harmless drinks? 77. How are wines manufactured? Describe the micro-organisms that start the fermentation. How are they introduced into the grape-juice? How is the fermentation arrested? What is cider? Does it contain alcohol? 78. What is perry? Koumyss? 79. What is the usual cause of effervescence? Is it always the result of fermentation? 80. What tendency has the habitual use of "light liquors"? Illustrate (foot-note). 81. What is the principle of distillation? Describe the process. What is "the worm of the still"? 82. What are the principal distilled liquors? How do they differ? What is their alcoholic strength? What imparts to them their peculiar flavor and odor? 83. What is the dangerous ingredient of all these drinks? Is alcohol decomposed by distillation?

CHAPTER III.—84. What is the use of the digestive apparatus? 85. What are the five stages in the preparation of food for the needs of the body? 86. Which of these stages is under control of the will? Why must all the stages be properly carried through? 87. Use of the senses of taste and smell. 88. Use of the teeth. Of the cheeks and tongue. What is the masseter muscle? What is said of the sound accompanying muscular contraction? 89. What is the saliva? What are the parotid glands? How does their secretion differ from that of the other salivary glands? Does the secretion of saliva vary in amount at different times? Illustrate (foot-note). 90. What are the properties of the saliva? Its effect upon starch? Its use in the preparation of food. Illustrate. How much saliva is secreted daily? What is said of the care of the teeth (note)?

CHAPTER IV.—91. What is the alimentary canal? 92. What is mucous membrane? Its minute structure? What are epithelial and epidermal cells (foot-note)? 93. Describe the character and arrangement of the muscles in the alimentary canal. What is the result of their contraction? 94. What is serous membrane? Its use? 95. What are the pharynx and œsophagus? How do we swallow? 96. How large is the stomach? What is its shape? Where are its two openings, and what are they called? What is the great pouch of the stomach? In what direction do substances pass through the stomach, and how is this regulated? 97. What portions of the food are digested in the stomach? 98. Describe the accident to St. Martin, and its result. 99. What is the appearance of the interior of a healthy stomach? 100. How does the gastric juice appear during secretion? When is it secreted? 101. What two necessary ingredients has the gastric juice? Can food be digested outside of the body? How much gastric juice is secreted daily? 102. What part do the muscles of the stomach perform during digestion? 103. What is the appearance of the interior of the stomach during indigestion? Is the gastric juice secreted at such times? How is the tongue affected? 104. How long a time is required for stomach-digestion? 105. How does thorough mastication assist digestion, and why? 106. What is said of eating too little? 107. Why is it harmful to eat too much? 108. Why should we not eat between meals? 109. What is hunger? Why is plain food the best? 110. How are we to know when we have eaten enough? How far apart should our meals be? 111. How are we to judge of what to eat? 112. What is said of pepper and mustard? What is the golden rule about eating?

QUESTIONS. 319

113. What is the natural drink? 114. What are the effects of tea and coffee? 115. What are the effects of pure alcohol upon the stomach? 116. Of diluted alcohol? Does it interfere with digestion? 117. Effect of an intoxicating dose of alcohol upon the stomach? 118. What serious changes may be brought about in the stomach by alcohol? 119. Effect of opium on the stomach? Of morphia? What bad effect may follow the injection of morphia under the skin (footnote)? 120. What are the effects of tobacco due to? How much nicotine in a cigar? 121. What is the effect of tobacco when first used? 122. Effect of tobacco on digestion? How does the use of tobacco lead to the use of alcohol? 123. Is opium a food? How does it sometimes seem to act as a food? 124. How is candy adulterated? What kinds are to be avoided? Why? 125. What is the trichina spiralis? The cysticercus cellulosæ? How can they be killed? What advantage in boiling drinking-water? What are the general rules about eating?

CHAPTER V.—126. What becomes of the fats, sugars, and starches in the stomach? What is the chyme? 127. What is the structure of the small intestine? How does it join the large intestine? What is the course of the latter? What is the appendix vermiformis (foot-note)? 128. How do the muscular fibers of the intestine contract? What is the result? 129. What is the duodenum? 130. What is the pancreas? What is its use? 131. What is the liver? Its secretion? 132. What other function has the liver? 133. What is the bile? Where is it discharged into the intestine? What reason is there for supposing that it is an excrementitious fluid? What is jaundice? What is the result if the bile is prevented from entering the intestine? Illustrate. Does the bile leave the body? What is the inference from these facts? 134. What are the intestinal juices? How are they useful? 135. How is it shown that food is absorbed from the intestine? What vessels absorb it? Origin of the name lacteal (foot-note)? 136. What is the peritonæum? Its use? How do the blood-vessels and lacteals get to the intestine? Where do they terminate? 137. What are the villi? Their size, number, and structure? 138. What are the lymphatics? The lymph? The lymphatic glands? What is the function of the lymphatics? What are the lacteals? The thoracic duct? 139. Where is the blood from the intestines carried? 140. What is the function of the villi? What is the chyle? How is it absorbed (foot-note)? 141. Describe the changes in the blood during digestion. 142. What is the effect of alcohol upon the liver? How is cirrhosis produced? 143. What are the general effects of alcohol upon nutrition? Of tobacco? 144. What is the spleen? Why would it seem to be an important organ? What is the result of its removal? What is probably its function (foot-note)?

CHAPTER VI.—145. What is the blood? Its physical properties? 146. What are the red blood-corpuscles? Their appearance and size? How do they differ in different animals? 147. What are the white blood-corpuscles? How numerous are they? What is their appearance and size? How much of the blood is made up of corpuscles? 148. What is the plasma? Its composition? 149. What is coagulation? What is the clot? The serum? Why is the clot red? How does serum differ from plasma? 150. What conditions affect the rapidity of coagulation? Will blood coagulate inside of the body? Illustrate. 151. What is the total amount of blood in the human body? 152. What gas is absorbed by the blood from the air? How important is it to life? Illustrate. 153. What process causes the blood to change color? 154. Effect of alcohol on the blood? Of tobacco?

CHAPTER VII.—155. Is respiration a simple process? 156. What is the structure of the interior of the nose? Why is it better to breathe through the nose than through the mouth? 157. What is the trachea? What are the bronchi? How are these tubes kept open? Do the rings of the trachea entirely surround it? Why not? What is the larynx? 158. What is the glottis? Where is it situated? What are the vocal chords? What is the epiglottis? Its use? 159. What are the lungs, their situation, appearance, and weight? 160. What is the minute structure of the lungs? What is a pulmonary lobule? A pulmonary vesicle? Their size? 161. What peculiarity has the lining membrane of the lungs? What is the function of the cilia? 162. Effect of alcohol on the nose and throat? Of tobacco? 163. On the lungs? Does alcohol prevent consumption? 164. What is asthma? 165. Where are the blood-vessels of the lungs situated? 166. What is the pleura? Its use? What is pleurisy? 167. How is inspiration effected? What are the situation and use of the diaphragm? What is hiccough (foot-note)? 168. How is expiration effected? 169. Which is the more powerful action, inspiration or expiration? 170. How much air is taken into the lungs with each inspiration? What is the entire capacity of the lungs? How much air can be expelled with a forcible expiration? 171. What is meant by the diffusion of gases? What is the function of the cilia? How is the air in the lungs changed otherwise than by respiration? 172. How much air is respired daily? How would you calculate the amount? 173. What is the composition of air as it enters the lungs? How is it changed there? How much water is expired daily? 174. How much oxygen disappears in the body? How was the carbon dioxide formerly supposed to be formed? How can this theory be disproved? 175. What gives the breath its peculiar odor? 176. How is the blood changed in the lungs? What is the cause of the change in color (foot-note)? 177. Where is the carbon dioxide found? How much is expired daily? 178. Effect of alcohol on the respiration? Of opium? Effect of opium on the secretions? 179. How is the air changed by respiration? Importance of moisture in the air (foot-note). 180. Effect of plants on the composition of the air. 181. How does the air of houses become unfit to breathe? What is the effect of combustion? What is the most dangerous impurity added to the air by the breath? 182. What is ventilation? How can it be effected? What becomes of the organic matter of the breath in the open air? 183. What are contagious diseases? How are they communicated from one person to another? Malarial fevers, sewer-air (foot-notes). 184. How are these diseases prevented from spreading? What precautions are to be taken in the sick-room? And why?

CHAPTER VIII.—185. What is asphyxia? How produced? How is the color of the surface of the body affected by it? 186. How long may a person remain under water without dying? 187. What should be done to resuscitate a person who has been under water until he is unconscious? And why? 188. What is the object of artificial respiration? Describe the method of performing it. What is the first sign of recovery? 189. What precautions are necessary with regard to the throat and tongue? What should be done when the person begins to breathe? Importance of fresh air (foot-note).

CHAPTER IX.—190. How is the circulation of the blood effected? 191. What is the heart? Its situation? What is the pericardium? 192. Describe the double circulation. 193. Describe the course of the blood in the two sides of the heart. Why does the blood always go forward and never backward? 194. How is each side of the heart divided? How many cavities are there altogether? What are they called? What is the course of the blood through them? 195. How many sets of valves has the heart? Where are they situated? What is their use? 196 What large blood-vessels are connected with the heart? 197. Describe the circulation of the blood more fully. 198. What peculiarity in the arrangement of certain valves in the heart? And what is the effect of this peculiarity? 199. How does the heart contract and relax? 200. What are the sounds of the heart? How are they produced? 201. How is the beating of the heart regulated? 202. How frequently does the heart contract? How may the rapidity of contraction vary, and what conditions affect it? Can its contrac-

tions ever be controlled by the will? Give illustration. 203. Effect
of alcohol upon the heart? How is a fatty heart produced by it?
204. Effect of tobacco on the heart? To what are the evil effects of
cigarette-smoking due?

CHAPTER X.—205. How is the work of the heart assisted? 206.
What is the structure of the arteries? How many layers? What is
the use of the elastic fibers? 207. How is the pulse produced? Where
can it be felt? 208. How do the arteries divide into branches, and how
do their walls change as they become smaller? 209. What are the
capillaries? How large are they? How numerous? Illustrate. 210.
What is the structure of the veins? What is the result of their structure? What are the venæ cavæ? 211. What effect has respiration on
the flow of blood in the veins? And how? 212. What effect has muscular contraction? And how? 213. What is the use of the valves in
the veins? How can this be illustrated? 214. How do the capillaries
assist the venous flow? 215. What is said of communicating blood-
vessels? 216. Give a brief review of the course of the circulation.
How rapidly does the blood flow in the arteries? How rapidly in the
capillaries? Describe the circulation as seen in the frog's foot. How
does the blood nourish the tissues? What is the lymph? 217. What
artery carries black blood, and what vein scarlet blood? And why?
218. How rapidly does the blood flow in the veins? Why is not the
rapidity of the blood-current the same in arteries and veins? 219. How
rapid is the general circulation? What experiment has been made to
determine this? How long does it take for all of the blood to pass
through the heart? How is this calculated? 220. What are the vaso-
motor nerves? How do they affect the circulation? 221. Effect of
alcohol upon the blood-vessels? 222. What diseases of the vessels
are produced by it? What is aneurism? How may apoplexy be
caused? 223. What permanent change in the features may be produced by alcohol? 224. What is the aorta? What is its course?
225. What are the femoral arteries? Their course? 226. What are
the brachial arteries? Their course? The radial and ulnar arteries?
227. What arteries supply the head and face? What is the course of
the carotid arteries? 228. Where are the veins usually situated?
What are the jugular veins? What other large and important veins are
there? How many pulmonary arteries? How many pulmonary veins?

CHAPTER XI.—229. What happens when the circulation of blood
in any part of the body is obstructed? Illustrate. How is the swelling
caused? What is dropsy? What happens when the obstruction is removed? 230. What is the result of disease of the valves of the heart?
How is shortness of breath produced? How is general dropsy pro-

duced by such disease? What is the result of disease of the aortic valves? What may result from inflammation of the edges of the valves? How may gangrene be caused? 231. How do physicians detect heart-disease? 232. Effect of coagulation on bleeding. 233. What are some of the conditions of coagulation? Illustrate. 234. How can the bleeding from an artery be distinguished from that from a vein? 235. How does Nature stop hæmorrhage? 236. How does cold arrest bleeding? 237. What are styptics? Name some of them. What is their effect on the blood? What is the objection to their use? 238. How is compression applied to a wound? Where should an artery be compressed? And a vein? 239. How does Nature arrest bleeding permanently? How do surgeons assist the process? What is the use of ligatures? 240. State briefly the means of arresting hæmorrhage. 241. How is bleeding from wounds of the limbs to be stopped? How is the knotted handkerchief to be used? 242. What is the cause of fainting, and what is to be done for it? 243. What does shortness of breath always indicate? Why does exercise cause it?

PART IV.

CHAPTER I.—244. What is said of the difficulty of investigating the nervous system? 245. What are the two great divisions of the nervous system? What are they called? What are their functions? 246. What are the two forms of nervous tissue? 247. What is the structure of the white substance? How large are nerve-fibers? What is the myelin? The axis-cylinder? What are supposed to be the functions of these different parts of the nerve-fiber? 248. What does the gray substance consist of? How large are nerve-cells? What is their structure? What relation is supposed to exist between the cells and the fibers? What is the function of the fibers? What of the cells? 249. What is the structure of a nerve? How do its constituent parts vary in different situations? What kind of nervous substance is most abundant? 250. What are nervous ganglia? 251. Illustrate the functions of sensation and motion as exhibited by nerves. What is reflex action? Why so called? 252. What is said of the rapidity of the nervous force? Give an illustration. 253. Do nerves become exhausted? How? 254. How can it be shown that nerves are mere conductors of force? 255. What are the reasons for believing that the gray substance originates force? How does the nerve-force differ from electricity? 256. What part of the nervous system supplies the voluntary muscles with nerves? And what part the involuntary ones? What part is under our own control?

CHAPTER II.—257. What is the structure of the sympathetic system? How large are the ganglia? Where are they situated? What is said about the solar plexus (foot-note)? 258. What is said of the rapidity of action of the sympathetic system? How is it illustrated in disease? 259. How is it illustrated by the pupil of the eye? 260. What is the effect of dividing a sympathetic nerve? How has it been shown on the ear? What do these facts indicate? 261. If the sympathetic nerve supplying a gland is divided, what effect is produced? How is "watering of the mouth" caused? 262. How is blushing produced? Why is it more evident in the cheeks? What facts show that it is regulated by the sympathetic nerves? How can the blush be prevented? 263. What are the vaso-motor nerves? Why so called? Are the cerebro-spinal and sympathetic systems entirely distinct from each other? 264. What system regulates the process of digestion? How may this process be affected by the emotions? 265. What effect has cold on the sympathetic nerves? How may serious disease be brought on by exposure? How do persons catch cold, and how may colds be prevented (foot-note)? 266. What are the signs of exhaustion of the sympathetic system? How is it to be relieved?

CHAPTER III.—267. Describe the spinal cord. How are the gray substance and the white substance arranged respectively? What are the fissures of the cord? 268. How many pairs of spinal nerves are there? What parts of the body do they supply? 269. What are the properties of the spinal nerves? 270. What two kinds of sensibility exist in these nerves? How is this shown? 271. What is the effect of dividing a spinal nerve? How can sensation and motion both be conveyed by the same nerve? 272. How many roots has a spinal nerve? How does the posterior root differ in appearance from the anterior? How are the spinal nerves formed, and how do they divide? 273. What effect is produced by dividing the anterior root of a spinal nerve? What does this experiment prove? 274. What effect is produced by dividing the posterior root of a spinal nerve? What does this experiment prove? 275. What do these experiments show with regard to the structure of spinal nerves? 276. How is the spinal cord connected with the brain? What is said of the crossing of fibers from one side to the other? 277. Why are sensations always referred to the extremity of the nerve? Give illustrations. 278. What is the use of the gray matter in the center of the spinal cord? Describe the experiment with the frog. What does it show? 279. How does Nature perform similar experiments for us upon men? Illustrate. 280. What are automatic actions? Illustrate.

CHAPTER IV.—281. What is the distinction between the brain and the cerebrum? What is said of the removal of the cerebrum? 282. What is the structure of the cerebrum? What are the convolutions? How is the gray matter distributed? Is the cerebrum sensitive to injury? 283. What is the function of the cerebrum? 284. Does intelligence seem to increase with the size of the brain? Illustrate. What is the weight of the human brain? 285. How much do human brains differ in weight? Illustrate. How would you account for a small brain in a clever man? 286. What effect do injuries of the brain have upon the mind? 287. What is the effect of softening of the cerebrum in man? 288. State briefly the evidence that the cerebrum is the seat of intelligence? 289. What is the structure of the cerebellum? Its size? 290. What is the effect of disease or injury of this organ in man? What is inferred from these facts as to the function of the cerebellum? 291. What is the function of the tubercula quadrigemina? How can this be proved? Where do the powers of sensation and motion reside? How is this shown? 292. Describe the situation and relations of the medulla oblongata. What is the vital knot? 293. How are the movements of respiration regulated by the medulla oblongata? 294. Illustrate the automatic action of this ganglion. What is the result of its destruction?

CHAPTER V.—295. How do the nerves of the brain differ from the spinal nerves? Which are the most important ones? What parts does the trigeminal nerve supply? What is its function? What parts are supplied by the facial nerve? What is its function? How may it be affected by exposure to cold? 296. What is the course of the sciatic nerve? What is the cause of the foot "being asleep"? 297. What is the importance of reflex action? Give illustrations (foot-note). 298. Why is education so important? Why are habits easily formed in the young? Why do they need the guidance of older persons? 299. Why is it important that the brain should have sufficient exercise? 300. What is the effect of over-exercise? 301. Why is sleep necessary? What takes place during sleep? Why do we wake up? What is the best time for sleep? Why? How much sleep is necessary? 302. What is said of abuse of the nervous system, and its danger? 303. Effect of a small amount of alcohol upon the nervous system? 304. Of a larger dose? How is double vision produced? 305. Effect of very large doses? After-effects? 306. How is alcohol supposed to cause these results? 307. What is chronic alcoholism? Its symptoms? 308. How is delirium tremens caused? Its symptoms? How may it end? 309. Effect of alcohol upon the moral nature? 310. Effect of alcohol on nervous tissue? 311. How may children suffer from the drinking habit of the parents? 312. What poisonous substances

22

besides alcohol are found in alcoholic drinks? Their effect upon the body? 313. Effect of tobacco on the nervous system? How does it produce its so-called soothing effect? What bad results may follow its use? How are they to be accounted for? 314. What marked difference in the effects of alcohol and opium? 315. Effect of a small dose of opium? 316. Of a larger dose? 317. Of a dangerously large dose? 318. How is the opium-habit formed? 319. Effects of opium on the mental, moral, and physical nature? 320. Effect of the chloroform-habit? 321. Of the chloral-habit? 322. Explain the cause of the craving due to the habitual use of narcotics.

PART V.

CHAPTER I.—323. In what points do the skin and mucous membrane resemble each other? What are the functions of the skin? 324. What is the structure of the skin? The epidermis? The derma? How thick is the skin? 325. How is the skin connected with the tissues beneath it? What are the papillæ? Their situation, size, and arrangement? 326. What causes the difference of complexion in different persons? What is an albino? What seems to be the use of the pigment-cells? 327. What are the nails? How do they grow? What is the use of them? 328. What are the hairs? How do they grow? What causes the difference in color of the hair in different persons? How much of the body is covered with hair? 329. What are the sebaceous glands? What is their structure? Where are they situated? What is their function? What is the result when their openings are obstructed? 330. What are the sweat-glands? Where are they situated? What is their structure? How large are they, and how numerous?

CHAPTER II.—331. What is the effect of pressure on the skin? What diseases may be produced by it? 332. What was Lavoisier's theory of the cause of animal heat? How overthrown? Liebig's theory? What objection is there to it? What is believed to be the true theory? 333. What is the normal temperature of the human body? How much may it vary? 334. What is the effect of a great fall or a great rise in the temperature of the body? What external temperatures are we often exposed to? 335. What is the effect of a low external temperature on the body? How is the appetite affected? How does the body accommodate itself to such a temperature? 336. What is the effect upon the body of a high external temperature? 337. How is the temperature of the body regulated by the perspiratory glands? 338. Does alcohol increase or diminish the power of endurance? Explain its effect in extreme cold and heat. Give illustrations. 339. What is the insensible perspiration ? What is its daily amount ? How may it be immensely increased? 340. What is the effect of exposure to dry heat ? 341. What is the effect of exposure to moist heat ? 342. What is said of respiration through the skin ? How much carbon dioxide is thrown off by it (foot-note)? What is said of absorption through the skin (foot-note)? 343. How does the skin become covered with impurities ? Why should these be removed ? 344. What is the effect of a cold bath ? How long should the bath last ? 345. What is the effect of a warm bath ? Why should exposure to cold be avoided after it ? 346. State the rules for bathing and the reasons for them ? 347. How should the face and hands be cleaned ? What is the objection to soap? How is " chapping" produced ? How may it be prevented ? What is the best soap? 348. What causes dandruff? How should the scalp be cleaned ? What is said about cleaning and cutting the hair ? 349. How should the nails be cut ? Why should they not be cut close to the flesh ? What causes " hangnails "? How may they be prevented? What causes the white spots on nails? 350. What is the object of clothing? What is the best material to wear next the skin ? Why? Why is linen bad for this purpose ? Why should the under-clothing be changed at night ? What is said of tight and high-heeled shoes ? What are the general rules for the selection of clothing ?

PART VI.

CHAPTER I.—351. Where does the sense of touch have its seat ? How do the nerves end in the papillæ of the skin ? 352. What do we learn by the sense of touch ? How does the sensitiveness of the skin vary in different parts of the body? How is this shown ? What is the most sensitive part (foot-note)? 353. Can the extremes of sensation be easily distinguished from each other? What is said of deception through the senses? Give illustrations (foot-note). 354. What relation do the special senses bear to the sense of touch ? What peculiarity is common to them ? 355. Where is the sense of taste located ? How are the nerves of taste distributed to the tongue ? How do they terminate ? How do we taste ? 356. How many modifications of the sense of taste are there ? What qualities are perceived in the mouth by the sense of feeling ? Is the aroma of a substance tasted or smelled ? How may this be determined ? 357. Why do substances have a different taste on the front and back of the tongue ? Give illustrations. Why does the sensation of taste persist for a time after the substance has been removed ? How does this affect the detection of delicate flavors ? Why do we smack our lips when tasting ? What is the use

of this sense ?   358. Where does the sense of smell reside ?   What is
the olfactory nerve ?   Its situation ?   How much of any substance is
required to affect the sense of smell?   Why do we sniff at odors in
order to appreciate them (foot-note) ?   Illustrate the occasional difficulty of distinguishing between the sense of smell and that of feeling.
359. Illustrate the acuteness of this sense in some men and lower animals.   What is the use of this sense ?

CHAPTER II.—360. What kind of impressions are perceived by the
sense of hearing ?   361. What are the three divisions of the ear?   What
is the use of the external ear ?   Is it movable ?   362. What is the tympanum ?   How do the air-waves reach it ?   What is the middle ear ?
Its situation and size ?   What is the Eustachian tube?   Its use?   363.
Where are the bones of the ear ?   How many are there, and how large?
What are they attached to, and how are they moved ?   364. What is
the structure of the internal ear ?   What are its chief parts, and what
is it filled with ?   How is the nerve of hearing distributed ?   365. How
is hearing effected ?   366. Upon what do the different qualities of
sound depend?   367. How do we determine the direction from which
sound comes ?   What is said of the function of the semi-circular canals
(foot-note)?   368. What is ventriloquism?   How are its effects produced ?   369. How is the ear protected from insects ?   How can they
be removed when they have entered the ear ?   What is said of the use
of ear-scoops ?   370. How may colds in the head affect the hearing ?
How may permanent deafness be caused ?   Influence of alcohol and
tobacco in causing deafness?

CHAPTER III.—371. How does the sense of sight differ greatly
from the other senses ?   372. What is light believed to be?   How
rapidly does it travel?   Describe the wave theory of light.   373. How
is the eye protected?   374. What is the shape of the eyeball?   Its
size ?   The three membranes ?   Their situation and structure?   375.
What is the vitreous humor?   What are *muscæ volitantes* (foot-note)?
The crystalline lens?   The iris?   The aqueous humor?   376. What
is the use of the sclerotic coat ?   Of the cornea?   Of the choroid ?
What is the cause of " dazzling " ?   What peculiarity have the eyes of
albinos (foot-note) ?   377. What is the structure of the retina ?   What
is the blind-spot ?   How may its existence be demonstrated ?   Why
does it not interfere with sight ?   378. What is the use of the lens?
379. What change takes place in the eye to enable us to see either
near or far objects at will ?   How is this change produced ?   What is
the ciliary muscle?   Its function?   380. What is the inner limit of
distinct vision ?   The farther limit ?   (In healthy eyes it is practically
infinitely distant.)   381. What causes the color of the iris ?   What are

the functions of the iris? What are spherical and chromatic aberration? How are they prevented by the iris? 382. How is it shown that the optic nerve conveys only the sensation of light? 383. Illustrate the persistence of impressions on the retina. 384. What is meant by "color-blindness"? 385. Explain near-sightedness and far-sightedness. 386. How is the eyeball moved? 387. What are the uses of the eyelids, eyebrows, and eyelashes? 388. What are the tears? Where do they come from, what is their use, and what becomes of them? How is weeping explained? How are the tears prevented from overflowing on to the cheeks? 389. Effect of excessive use of tobacco upon the eyes? What is tobacco-amaurosis? 390. What is said of the care of the eyes? Of their delicacy? Why is it bad to strain them? Why should they not be opened under cold water (foot-note)? What abuses of the eye should especially be avoided?

PART VII.

391. What is the larynx? Its structure, situation, and shape? 392. What are the vocal chords? How are they moved (foot-note)? 393. How is the voice produced? How is it modulated? 394. How is the larynx commonly abused? What effect has inflammation of the vocal chords upon the voice? How is "breaking" of the voice produced? 395. Why is rest necessary when the vocal chords are inflamed?

GLOSSARY.

Abdo'men. That portion of the trunk of the body which is situated between the diaphragm and the pelvis. The tissues surrounding and inclosing it, the muscles, skin, bones, etc., are called the *abdominal walls.*

Accommodation. The adjustment of the internal parts of the eye in correspondence with the varying distance of external objects, so that their images always fall upon the retina.

Albino. (Portuguese, from Latin, *albus*, white.) An animal in which pigment-cells are lacking, so that the hair and skin are white, and the pupil of the eye red.

Albu'men. (A Latin word, signifying the *white of an egg*, which is almost pure albumen.) A very nutritious substance of very complex chemical composition, constituting a large part of meats and some vegetables, and classed with the nitrogenous foods or proteids (q. v.).

Alcohol. (From the Arabic.) The common name of what chemists call ethyl-alcohol. Produced, together with carbon dioxide, by the fermentation of liquids containing sugar. Is present in all fermented fluids, the distilled liquors containing from 50 to 60 per cent; heavy wines, port, sherry, etc., 16 to 25 per cent; clarets and hocks, 7 to 16 per cent; and malt liquors, 2 to 10 per cent.

Aliment'ary Canal. The long tube extending from the mouth downward through the body, by which food is received, digested, absorbed, and the residue expelled. It includes the mouth, pharynx, œsophagus, stomach, small intestine, and large intestine.

Anatomical Element. A portion of a living body so small that it can not be further divided without the destruction of its organization.

Anatomy. (Greek, ἀνατομή, a cutting up.) The science that describes the appearance, structure, and situation of the different parts of the body.

Aor'ta. (Greek, ἀορτή.) The large artery which first receives the blood from the left ventricle of the heart.

Apoplexy. (Greek, ἀποπληξία, a sudden stroke.) The escape of blood from a broken blood-vessel into the substance of the brain.

Apparatus. A number of organs, differing in size and structure, which work together for the accomplishment of a particular object, e. g., the digestive apparatus, comprising the stomach, liver, pancreas, intestines, etc.

Appen'dix Vermifor'mis. (Latin, meaning the *worm-like appendage*.) A small projection from the cæcum (q. v.), about six inches long and half an inch wide, of the same general structure as the intestine, situated in the right groin, and of no apparent use. Small objects, such as cherry-pits, grape-seeds, etc., sometimes lodge in it, and give rise to a fatal inflammation.

A'queous Hu'mor. (Latin, *aqua*, water.) A fluid in the eye lying between the crystalline lens and the cornea, consisting almost entirely of water, with a little salt.

Ar'tery. (Greek, ἀρτηρία, from ἀήρ, air, and τηρεῖν, to contain, because, being always found empty after death, they were supposed to contain air during life.) A blood-vessel carrying blood away from the heart.

Asphyx'ia. (Greek, ἀσφυξία, pulselessness.) Suffocation.

Asth'ma. (Greek, ἄσθμα, a gasping or panting.) A spasmodic affection of the lungs, in which the caliber of the small bronchial tubes is diminished so much as to interfere with free respiration.

Au'ditory Nerve. The nerve of hearing.

Au'ricles. (Latin, *auris*, the ear ; called auricles from their fancied resemblance to ears.) Two cavities of the heart into which the blood first enters as it returns from the rest of the body.

Axis-Cylinder. A transparent or finely granular thread running in the center of every nerve-fiber, and believed to be the part along which the nerve-current passes.

Bacte'ria. (Greek, βακτήριον, a staff.) Small organisms, only visible under the microscope, shaped like rods, and believed to be the cause of many diseases.

Bi'ceps. (Latin, *biceps*, two-headed.) The large muscle extending on the front of the arm from the shoulder to the elbow. So called because it is attached to two different points of the shoulder, i. e., by two heads.

Blood-Corpuscles. The microscopic bodies which float in vast numbers in the blood, and are the most essential constituent of it.

Brach'ial. (Latin, *brachium*, the arm.) Pertaining to the upper arm, as the brachial artery.

Bron'chi. (Latin.) The tubes into which the trachea or windpipe is divided, and which convey air to the interior of the lungs.

Bronchi'tis. Inflammation of the mucous membrane lining the bronchi.

Cæ'cum. (Latin, *cæcus*, blind.) The beginning of the large intestine. So called because it bulges downward from the point where the small intestine enters it, and forms a sort of *cul-de-sac*, or blind pouch.

Cap'illaries. (Latin, *capillus*, a hair.) The microscopic blood-vessels through which the blood passes from the arteries to the veins.

Carbon Dioxide. Often called carbonic acid. A gas containing one part of carbon to two of oxygen. When inhaled it produces insensibility, followed by death if the inhalation is continued.

Carbona'ceous Foods. Foods containing a large proportion of carbon but no nitrogen. They also contain hydrogen and oxygen ; and, if the latter constituents are in the same proportion as in water, e. g., in starches and sugars, the foods are called carbohydrates ; otherwise, hydrocarbons, e. g., the oils and other fats.

Car'diac Orifice. (Greek, καρδία, the heart.) That opening of the stomach by which food enters. So called because it is nearest the heart.

Carot'id Arteries. (Greek, καρωτίδες, same meaning.) The large arteries supplying the front of the head and face.

Car'tilage. (Latin, *cartilago*, same meaning.) A bluish-white, elastic substance, found in various situations in the body, but especially covering the ends of bones which form joints.

Ca'sein. (Latin, *caseus*, a cheese.) A nitrogenous substance or proteid (q. v.), forming the chief part of the curd of milk, often pressed into cheese.

Catarrh. (Greek, κατάῤῥοος, a morbid discharge.) An acute or chronic inflammation of a mucous membrane, usually attended by an increased secretion of mucus.

Cau'da Equi'na. (Latin, meaning a *horse's tail.*) The lower end of the spinal cord, so called because it is split up into strings of nerves resembling somewhat a tassel or a horse's tail.

Cerebel'lum. (Latin, signifying a *little brain.*) A portion of the brain, constituting a distinct ganglion, situated under the posterior part of the cerebrum.

Cer'ebro-spinal System. The brain and spinal cord, with the nerves arising from them.

Cer'ebrum. (Latin, meaning *the brain.*) The larger and upper part of the nervous mass lying within the skull, and conceded to be the seat of the intellect.

Chlo'ral. A narcotic poison, formed from chlorine and alcohol. Used internally to produce sleep.

Chloroform. A narcotic poison produced by the distillation of chloride of lime and alcohol, and generally used by inhalation. It produces anæsthesia, with complete unconsciousness, and is therefore useful in surgical operations.

Cho'roid Coat. (Greek, χορίον, leather, because of its dark, leather-like color.) The middle coat of the eye, containing numerous pigment-cells and blood-vessels.

Chromat'ic Aberration. (Greek, χρῶμα, color.) The splitting up of white light into colored light as it passes through a lens, in consequence of the difference in the refrangibility of its components. The image produced in the focus of such a lens will be fringed with color. The violet rays, being the most rapid in their undulations, are brought to a focus first, and the red rays, being the slowest, last, so that if a screen be placed in the focus of the violet rays, when a lens is not achromatic, the image will be fringed by red rays, which have not yet converged to a focus ; and if the screen be put in the focus of the red rays, the image will be fringed by violet rays, which have passed their focus, and begun to diverge.

Chyle. (Greek, χυλός, same meaning.) The whitish fluid that is absorbed from the intestine into the lacteals.

Chyme. (Greek, χυμός, juice.) The grayish, pulpy fluid into which food is converted in the stomach.

Cil'ia. (Latin, *cilium*, an eyelash.) Hair-like projections on the free extremity of certain epithelial cells.

Cil'iary Muscle. A small, circular muscle in the interior of the eye-ball, which assists in the function of accommodation.

Ciliated Epithelium. Epithelium having cilia on the free extremities of the cells.

Coch'lea. (Greek, κοχλίας, a spiral shell.) A portion of the internal ear, shaped like the interior of a snail-shell.

Com'press. A pad or bandage applied directly to a wound, in order to compress it.

Connective Tissue. A delicate lace-work of fibrous threads which extends through all organs of the body, binding their elements together.

Convolutions of the Brain. The projections on the surface of the brain, which look somewhat like those of a walnut-meat.

Co-ordination. The bringing of several different things into such relations with one another that they perform their functions in harmony.

Cor'nea. (Latin, *cor'nu*, a horn.) The transparent portion of the external coat of the eye.

Crys'talline Lens. The lens of the eye, by which light is brought to a focus on the retina.

Cutis. (Latin, meaning *the skin*.) The true skin, containing the hair-bulbs, blood-vessels, nerves, etc. The hard, insensitive layer outside of the cutis is called the cuticle, i. e., the little skin.

Cysticer'cus Cellulo'sæ. (Part Latin and part Greek, κύστις, a bladder, κέρκος, a tail, and *cellula*, meaning the bladder-tailed animal of the cellular [or connective] tissue.) A parasite, found in measly pork. When eaten by a human being, it is developed into a tape-worm in the intestine.

Der'ma. (Greek, δέρμα, the skin.) Same as *cutis* (q. v.).

Diaphragm. (Greek, διάφραγμα, a partition.) The muscular partition which separates the thorax, or chest, above, from the abdomen below.

Disinfectants. Drugs which destroy the germs or particles of living matter that are believed to be the causes of various diseases.

Dislocation. The displacement of the end of a bone.

Drum of Ear. The membrane which separates the middle ear from the external ear, and on which the air-waves (sound-waves) first strike.

Duode'num. (Latin, *duodeni*, twelve each.) The first eight or ten inches of the small intestine nearest the stomach. So called because its length is about twelve fingers' breadth. It differs somewhat in structure from the remainder of the intestines, and is a little larger.

Epider'mal Cells. The cells composing the epidermis.

Epider'mis. (Greek, ἐπί, upon, and δέρμα, the skin.) The same as *cuticle* (q. v.).

Epiglot'tis. (Greek, ἐπί, upon, and γλωττίς, the glottis.) A leaf-shaped cartilage, covered with mucous membrane, situated at the base of the tongue, just above the glottis.

Epithe'lium. (Greek, ἐπί, upon, and θηλή, a nipple.) A general term applied to various kinds of cells which cover all the free surfaces of the body, and are called epithe'lial cells.

Ether. (Greek, αἰθήρ, the pure upper air.) A narcotic poison, made by distilling sulphuric acid with alcohol, and used as an anæsthetic in surgical operations.

Ether, The. A very subtile form of matter, which is supposed to fill all space, permeating all bodies, to the vibrations of which the phenomena of light and heat, electricity and magnetism, are believed to be due.

Eusta'chian Tube. (From Eustachi, its discoverer.) The tube or canal leading from the middle ear to the throat.

External Auditory Canal. The canal or tube leading from the external ear inward to the drum.

Facial Nerve. The motor nerve of the face.

Fem'oral Artery. (Latin, *femur*, the thigh.) The main artery of the thigh.

Fiber. A slender, microscopic thread of nitrogenous matter, very strong, tough, and elastic. When these threads are delicate, and loosely woven, they form connective tissue. When larger and more closely woven, and with a tendency to a longitudinal arrangement, they make fibrous tissue. There are other differences between these tissues, but they are too technical for insertion here.

Fi'brin. A nitrogenous material contained in the flesh of animals, and classed among the proteids. It is also produced by the coagulation of blood, constituting the greater portion of the clot.

Fol'licles. (Latin, *folliculus*, a little sack or pouch.) A name applied to little pits or depressions in the skin or mucous membrane, out of which is poured a secretion, differing in different situations.

Frontal Si'nuses. Cavities in the bones of the skull, just over the eyebrows, the front wall causing the projection noticeable there.

Function. (Latin, *functio*, from *fungor*, to perform.) The work performed by any organ of the body is called its function.

Ganglion. (Greek, γάγγλιον, a tumor.) A name applied to certain masses of gray nerve-matter, which have independent functions, and are separated from other masses of gray matter by intervening white matter.

Gangrene. (Greek, γάγγραινα, gangrene.) Death of a portion of the soft parts of the body, while still attached to the living tissues. Generally caused by deprivation of blood.

Gastric. (Greek, γαστήρ, the belly.) Pertaining to the stomach.

Gland. (Latin, *glans*, an acorn.) An organ which separates something from or adds something to the blood that passes through it, by its own activity.

Glot'tis. (Greek, γλωττίς, the glottis.) The upper opening of the larynx.

Glucose. (Greek, γλυκύς, sweet.) A kind of sugar, also called grape-sugar, found in fruits in considerable quantity. Produced artificially by the action of a mineral acid on starch. Cane-sugar is converted into glucose in the stomach.

Graduated Compress. Several pads of soft material, of different sizes, placed upon each other with the smallest underneath and the largest on top. Used for compression of wounds.

Granular Matter. Microscopic particles of matter, generally of a nitrogenous nature, found in great quantity in certain tissues in the form of fine granules, whose precise character and function are not yet determined.

Gray matter. The ash-colored portion of the nervous system, consisting mainly of nerve-cells, and believed to originate nerve-force.

Great Pouch of the Stomach. The enlargement of the stomach toward the left of the cardiac orifice.

Hæmoglobin. (Greek, αἷμα, blood, and Latin, *globus*, a sphere.) The coloring-matter of the blood, consisting of a proteid substance in combination with iron, and making up the greater part of the blood-corpuscles.

Homol'ogous Bones. Those bearing similar relations to adjacent parts, and having similar special functions.

Hygiene. (Greek, ὑγιεινός, healthy.) The science which describes what is healthful and what is injurious to the body, and how health may be maintained.

Insensible Perspiration. Perspiration which is too small in quantity to collect in drops, and is therefore not ordinarily perceived.

Internal Ear. That part of the ear which lies imbedded in bone, and in which the branches of the nerve of hearing are distributed.

Involuntary Muscles. Those that are not under the control of the will.

Iris. (Greek, ἶρις, a rainbow.) The colored muscular curtain in the front of the eyeball which surrounds the pupil; so called on account of its variety of color.

Jaundice. A disease, in which the coloring-matter of the bile is found in the blood and stains all the tissues yellow.

Jug'ular Veins. (Latin, *jugulum*, the neck.) The large veins in the neck which return the blood from the head and face.

Labyrinth. A portion of the internal ear, so called from its winding passages.

Lach'rymal. (Latin, *lachryma*, a tear.) Pertaining to the tears, as the lachrymal gland, which secretes them.

Lac'teals. (Latin, *lac, lactis*, milk.) Those vessels of the lymphatic system which absorb the chyle from the small intestine and appear like white threads and cords during digestion.

Lar'ynx. (Greek, λάρυγξ, the larynx.) The organ of voice, situated at the top of the windpipe, producing the protuberance in the neck commonly known as Adam's apple.

Latis'simus Dor'si. (Latin, meaning the *broadest of the back*.) A large, triangular muscle, attached throughout one side to the spine, and at the opposite angle to the arm near the shoulder, the fibers running from the spine to the arm.

Lens. (Latin, *lens*, a lentil.) A transparent substance bounded by curved surfaces, so that light passing through it is refracted, i. e., bent out of its course. So called from its shape.

Lig'ament. (Latin, *ligamentum*, a band.) A closely knit mass of fibrous tissue, white, glistening, tough, and inelastic, which binds bones together at the joints.

Lig'ature. (Latin, *ligatura*, a band.) A cord, used in surgery for tying blood-vessels, so as to prevent or stop bleeding.

Lymph. (Latin, *lympha*, water. Poetic.) A yellowish, transparent, saltish fluid, found in the lymphatic vessels.

Lymphat'ic Vessels. A system of tubes, varying in size, found in all parts of the body, like the blood-vessels, possessing great absorbent powers, containing always the lymph, which they collect and pour into the large veins near the heart.

Lymphat'ic Glands. Small bodies, varying in size from a hemp-seed to an almond, scattered at intervals along the lymphatic vessels, which enter them and emerge again. Their functions are imperfectly known.

Marrow. The reddish, pulpy, fatty substance contained in the shaft of the long bones.

Mas'seter Muscle. (Greek, μασητήρ, a chewer.) A square muscle, attached to the skull and to the horizontal part of the lower jaw. Used in bringing the jaws together.

Medul'la Oblonga'ta. (Latin, meaning the *pith or marrow lengthened out*.) An enlargement at the upper extremity of the spinal cord, just as it enters the skull, but before its fibers diverge to the different parts of the brain. One of its chief functions is the control of respiration.

Middle Ear. A small cavity, containing the minute bones which serve to conduct the sound-waves from the drum of the ear to the internal ear.

Molec'ular. Pertaining to molecules, i. e., the smallest particles of matter into which any substance can be divided without becoming something else.

Motor. Producing or giving rise to motion. A motor nerve is one whose function is to excite motion, instead of conveying sensation.

Mucous Membrane. The soft, velvety membrane which lines all those cavities of the body which communicate with the external air. It is continuous with the skin, and secretes a glairy fluid, which keeps its surface constantly moist.

Mus'cæ Volitan'tes. (Latin, meaning *flitting flies.*) Little strings or fibers, which float in the interior of the eye, and which can be readily seen by looking at a white wall, appearing like strings of beads, or transparent spheres clustered together, and moving when the eye moves. Found in healthy eyes. Spots which remain fixed imply disease.

Myelin. (Greek, μυελός, marrow.) The white nerve-matter which surrounds the axis-cylinder (q. v.), and forms the bulk of the nerve-fiber.

Narcot'ic Poisons. (Greek, ναρκόω, to benumb.) Poisons such as opium, alcohol, ether, and chloroform, which produce at first, or in small quantity, a feeling of exhilaration, followed, if more be taken, by insensibility and death, or, if the person gets well, by a period of depression, from which the system is slow to recover.

Nasal. (Latin, *nasus*, the nose.) Pertaining to the nose.

Nerve-Center. A mass of gray nerve-substance, believed to originate the nerve-force.

Nitrog'enous Foods. Substances, chiefly of animal origin, which contain nitrogen, in addition to carbon, hydrogen, and oxygen. Many of them also contain a little sulphur.

Non-nitrogenous. (See **Carbonaceous**.)

Nucle'olus. Diminutive of **nucleus**. A small, bright particle of matter inside of a nucleus, and forming a part of it.

Nu'cleus. (Latin, *nux, nucis*, a nut.) A small body contained in the interior of a cell, and forming a part of it.

Nystag'mus. (Greek, νυσταγμός, a nodding.) An affection of the eye, in which the eyeball oscillates constantly from side to side.

Œsoph'agus. (Greek, οἰσοφάγος, the gullet.) The tube, about nine inches long, that conveys substances from the throat to the stomach.

Olfact'ory. (Latin, *olfacio*, to smell.) Pertaining to the sense of smell, as the olfactory nerve.

Opium. A narcotic drug, obtained from the juice of the poppy.

Optic. (Greek, ὀπτικός, belonging to sight.) Pertaining to the sense of sight, as the optic nerve.

Orbit. The bony hollow in which the eyeball rests.

Osmo'sis. (Greek, ὠσμός, a thrusting or pushing.) The passage of substances through animal membranes, and some other media, in consequence of an obscurely understood power of selection in the medium by which matters in solution and fluids are enabled to go through in one direction rather than another, and some substances to pass rather than others. By a skillful direction of this process, matters in solution may be separated from each other. Osmosis is constantly going on in the living body, and by its aid the nutritive processes are carried on.   The passage of matter through a membrane into a vessel is called *endosmosis*, and the passage out,*exosmosis*.

Ossifica'tion. (Latin, *ossificatio*, from *os*, a bone, and *facio*, I make.) The change of any substance into bone.

Oxida'tion. The chemical union of oxygen with any substance, forming an oxide; e. g., combustion is a rapid union of carbon and oxygen, forming carbon dioxide.

Oxyhæmoglo'bin. (First part of the word oxygen, and hæmoglobin, q. v.)   Hæmoglobin loosely combined with oxygen.

Pan'creas. (Greek, πάγκρεας, the sweet-bread, from πᾶν, all, and κρέας, flesh.)   The organ whose secretion, called the pancreatic juice, digests fatty matters.   Known to butchers as the sweet-bread.

Papil'la. (Latin, *papilla*, a nipple.)   A minute projection from the outer surface of the derma, containing the terminations of nerves and blood-vessels.

Paraple'gia. (Greek, παραπληγία, paralysis.)   Paralysis of the lower part of the body on both sides.

Par'asite. (Greek, παράσιτος, a table-companion.)   Any living thing, animal or vegetable, whose nature it is to attach itself to some other living thing, and derive its nourishment from it.

Parot'id Glands. (Greek, παρωτίδες, the parotid glands, from παρά, beside, and οὖς, the ear.)   The large salivary glands situated just in front of the ears.

Pepsin. (Greek, πέπτω, to cook.)   The organic constituent of the gastric juice, one of the main agents in digestion.

Pericar'dium. (Greek, περί, around, and καρδία, the heart.)   The serous envelope of the heart.

Perios'teum. (Greek, περί, around, and οστέον, a bone)   The membrane, tough and fibrous, which envelops and clings closely to bone.

Peritonæ'um. (Greek, περιτόναιον, the peritonæum, from περί, around, and τείνω, to stretch.)   The serous membrane which lines the abdominal walls and envelops the organs contained in the cavity of the abdomen.

Phantom Limbs. A term applied by medical men to the sensory images produced by certain changes in the stumps of amputated limbs, which are of such a nature that the persons appear to themselves to feel pain, tingling, or other sensations in that part of the limb which has been separated from the body.

Phar'ynx. (Greek, φάρυγξ, the throat.)   That part of the throat which can be seen through the open mouth, without the aid of instruments.

Physiol'ogy. . The science that describes the functions of the different parts of living beings.

Pigment-Cells. (Latin, *pigmentum*, paint.) Cells naturally contain-ing coloring matter.

Plas'ma. (Greek, πλάσμα, formed matter.) The fluid portion of the blood before coagulation.

Pleu'ra. (Greek, πλευρά, the pleura.) The serous membrane which lines the chest and envelops the lungs.

Por'tal Vein. (Latin, *porta*, a gate.) The large vein which carries the blood from the digestive organs to the liver.

Pro'teids. (Greek, πρῶτος, first.) A class of substances, e. g., albu-men, fibrin, casein, etc., which contain carbon, hydrogen, oxygen, nitrogen, and sulphur, whose molecules are very complex, and which coagulate under proper conditions. So called because they form the basis and the greater part of animal bodies.

Pro'toplasm. (Greek, πρῶτος, first, and πλάσμα, formed matter.) An albuminous, somewhat jelly-like substance, which forms the princi-pal part of living beings. It is very complex in structure, combin-ing several of the proteids, probably with carbohydrates and fats. It is the only matter in the world that presents the phenomena of life.

Pul'monary. (Latin, *pulmo*, a lung.) Pertaining to a lung.

Pulmonary Artery. The blood-vessel that conveys blood from the heart to the lungs.

Pulmonary Lobules. The smallest independent divisions of the lungs, situated at the ends of the smallest bronchial tubes.

Pulmonary Veins. The blood-vessels that convey blood from the lungs to the heart.

Pulmonary Vesicles. The minute subdivisions of the pulmonary lobules.

Pulse. The blow given by an artery when suddenly distended. Generally felt with the finger on the radial side of the wrist.

Pupil. The circular hole in the iris through which light enters the eye.

Pylo'rus. (Greek, πυλωρός, a gate-keeper.) The opening by which food passes from the stomach into the intestine. Also called the pylor'ic orifice.

Ra'dial Artery. The artery lying on the radial or thumb side of the fore-arm.

Reflex Action. Motor action of a nerve-center in response to a stim-ulus without the co-operation of consciousness.

Ret'ina. (Latin, *retina*, the retina, from *rete*, a net, from the net-like or mesh-like expansion of the optic nerve.) The internal nervous coat of the eye, on which images of external objects must be formed in order to make sight possible.

Rods and Cones. A term applied, on account of their shape, to cer-

23

tain nervous structures in the outer layer of the retina farthest from the light. They are packed side by side, like the seeds of a sunflower, and are believed to be the structures by which the ether-waves are converted into a nerve-current, which, on reaching the brain, gives us the sensation of light.

Sali'va. (Latin.) The spit or spittle, composed of fluids secreted by the mouth.

Sal'ivary Glands. The glands that secrete the saliva.

Sarto'rius. (Latin, *sartor*, a tailor.) A long, ribbon-like muscle extending from the outer and upper part of the hip-bone to the leg, just below and on the inner side of the knee. When it contracts it raises the leg and turns it inward, as in crossing the legs.

Sciat'ic Nerve. (Contracted from *ischiatic*, Greek, ἰσχιάς, pain in the leg.) The largest nerve of the lower limb, extending from the pelvis to the foot at the back of the leg.

Sclerot'ic Coat. (Greek, σκληρός, hard.) The outer coat of the eye, consisting of white fibrous tissue. The white of the eye.

Seba'ceous Glands. (Latin, *sebaceus*, from *sebum*, tallow.) Certain glands in the skin, generally situated near the roots of hairs, secreting a whitish, fatty, tallow-like substance, which tends to keep the skin and hairs soft and pliable.

Semicircular Canals. A part of the internal ear filled with fluid, and containing filaments of the auditory nerve. So called from their shape.

Serous Membrane. A delicate kind of membrane secreting a small amount of fluid, and lining all cavities in the body which do not communicate with the external air, excepting the joints.

Se'rum. That fluid which remains after the clot has been separated from coagulated blood.

Sewer-Air. A mixture of gases formed by the decomposition of organic matters in sewers. Its most dangerous constituents are the particles of solid matter and the living microscopic organisms that float in it.

Spher'ical Aberration. The unequal refraction of light by a lens, at different distances from its center, in consequence of which the rays are not brought to the same focus, and the image is blurred and indistinct.

Spinal Canal. The long cavity inside the backbone, in which the spinal cord lies.

Spinal Cord. The long mass of nerve-substance that lies inside the spine.

Spinal Nerves. The nerves that take their origin from the spinal cord.

Spine. The backbone.

Sprain. Rupture of a ligament, or of fibers thereof, by the sudden twisting of a joint.

Stape'dius. (Latin, *stapes*, a stirrup.) A small muscle of the middle ear. So called because it is connected with the *stapes*, one of the bones of the ear.

Stria'tion. (Latin, *striæ*, lines or markings.) The parallel marking of voluntary muscular fibers at right angles to their length, only to be seen with the microscope.

Styp'tics. (Greek, στυπτικός, astringent.) Substances that produce shrinking or contraction of living tissues. Astringents.

Su'tures. (Latin, *sutura*, a seam.) The joints between the jagged edges of the bones of the skull, which appear as if dove-tailed.

Sympathetic System. That portion of the nervous system that controls the functions of organic life, i. e., the functions that are or should be performed unconsciously.

Syno'vial Membrane. (Greek, σύν, with, and Latin, *ovum*, an egg.) The membrane which lines the joints. So called because its secretion, the synovial fluid, resembles the white of an egg.

System. Several different organs, of similar structure, spread throughout the body, and performing similar functions.

Tape-worm. A worm whose larvæ are introduced with the food, and developed in the intestine. It is composed of many segments attached to each other, and may grow to an enormous length, twenty or thirty feet.

Tendons. The fibrous cords by which the muscles are united with the bones.

Thorac'ic Duct. The tube which collects the lymph from the lacteals and most of the other lymphatic vessels, and discharges it into the veins near the heart.

Tho'rax. (Greek, θώραξ, the chest.) The part of the trunk situated above the diaphragm. The chest.

Tissue. Two or more anatomical elements united so as to form one structure.

Tra'chea. (Greek, τραχεῖα, rough.) The windpipe.

Trichi'na Spira'lis. (Greek, τρίχινος, made of hair, and Latin, *spiralis*, spiral.) A parasitic worm often found in the flesh of the hog, and sometimes introduced with raw ham or pork into the human body, where it produces serious and even fatal disease. So called because it is slender in form, and is usually seen coiled up in a spiral.

Trigem'inal Nerve. (Latin, *trigemini*, three of a kind.) The nerve of sensation of the face. So called because there are three large branches of it.

Tuber'cular Quadrigem'ina. (Latin, *tubercula*, small swellings, and *quadrigemina*, four of a kind.) Certain protuberances, four in number, near the base of the brain—the nerve-centers of vision.

Tym'panum. (Greek, τύμπανον, a drum, or Latin, *tympanum*.) The cavity of the middle ear.

Ulnar Artery. The artery lying on the inner or little-finger side of the fore-arm.

U'rea. An excrementitious substance expelled from the body through the kidneys.

Vaso-motor Nerves. (Latin, *vasa*, vessels.) Nerves which control the movements of the muscles lying in the walls of the blood-vessels.

Veins. Blood-vessels which convey blood toward the heart.

Ven'æ Ca'væ. (Latin, meaning *the hollow veins*.) The large veins which collect blood from the smaller ones and discharge it into the heart.

Ven'tricles. (Latin, *ventriculus*, a little stomach.) The two largest cavities of the heart, from which the blood is sent out into the arteries.

Ver'tebra. (Latin, *vertebra*, a vertebra, from *verto*, I turn.) One of the smaller bones of which the spine is composed.

Ver'tebral Arteries. Arteries which pass up into the head through the vertebræ of the neck.

Vestibule. A part of the internal ear.

Vil'li. (Latin, meaning *closely-set hairs*.) Minute projections from the internal surface of the small intestine, containing blood-vessels and lacteals, by which the nutritious portions of the food are absorbed.

Vital Knot. A portion of the medulla oblongata (q. v.), the destruction of which kills instantly.

Vit'reous Humor. (Latin, *vitreus*, glassy.) A transparent, jelly-like mass filling the space in the eyeball between the lens and the retina.

Vocal Chords. Two pearly-white, glistening, fibrous bands at the upper part of the larynx, whose vibration causes the voice. The opening between them is the glottis (q. v.).

Voluntary Muscles. Muscles that are under control of the will.

White Matter. A kind of nervous tissue composed mainly of fibers, and serving to conduct nerve-force, but not to originate it.

INDEX.

THE END.

Approved Text-Books

For Temperance Instruction in Physiology and Hygiene.

AUTHORIZED PHYSIOLOGY SERIES.

I. HEALTH FOR LITTLE FOLKS . . 30 cents
II. LESSONS IN HYGIENE (Johonnot and Bouton) . . . . . . 45 cents
III. OUTLINES OF ANATOMY, PHYSIOLOGY AND HYGIENE (R. S. Tracy) . . $1.00

ECLECTIC TEMPERANCE SERIES.

I. THE HOUSE I LIVE IN . . . 30 cents
II. YOUTHS' TEMPERANCE MANUAL . 40 cents
III. GUIDE TO HEALTH . . . . 60 cents

PATHFINDER SERIES.

I. CHILD'S HEALTH PRIMER . . . 30 cents
II. YOUNG PEOPLES' PHYSIOLOGY . . 50 cents
III. HYGIENIC PHYSIOLOGY (J. D. Steele). $1.00

UNION SERIES.

I. PHYSIOLOGY AND HEALTH NO. 1 (primary) . . . . . 24 cents
II. PHYSIOLOGY AND HEALTH NO. 2 (intermediate) . . . . 30 cents
III. PHYSIOLOGY AND HEALTH NO. 3 (advanced) . . . . . 50 cents

All of the above series have been prepared under the personal supervision of Mrs. Mary H. Hunt, National and International Superintendent, Department of Scientific Instruction of the Woman's Christian Temperance Union, and each book bears the official indorsement of the Union.

Copies of the above books will be mailed, postpaid, on receipt of price. Full price-list will be sent on application. Correspondence is cordially invited.

AMERICAN BOOK COMPANY.

NEW YORK .·. CINCINNATI .·. CHICAGO.

[*13]

General Science.

Doerner's Treasury of Knowledge.

By CELIA DOERNER. { Part I. . . . . $0.50.
Part II. . . . . .65.

This book is designed to fill a gap in the ordinary course of instruction, and furnishes in a small compass much useful and important information. Since it combines entertainment with instruction, it will be found especially useful to parents as an addition to the child's home library.

Hooker's Child's Book of Nature. (COMPLETE.)

By WORTHINGTON HOOKER, M. D., ... $1.00.

Three parts in one: Part I. Plants; Part II. Animals; Part III. Air, Water, Heat, Light, etc. Designed to aid mothers and teachers in training children in the observation of Nature. It presents a general survey of the kingdom of Nature in a manner calculated to attract the attention of the child, and at the same time to furnish him with accurate and important scientific information.

Monteith's Easy Lessons in Popular Science.

By JAMES MONTEITH. . . . . $0.75.

This book combines the conversational, catechetical, blackboard, and object plans, with maps, illustrations, and lessons in drawing, spelling, and composition. The subjects are presented in a simple and effective style, such as would be adopted by a good teacher on an excursion with a class.

Monteith's Popular Science Reader.

By JAMES MONTEITH. . . . . $0.75.

This contains lessons and selections in Natural Philosophy, Botany, and Natural History, with blackboard, drawing, and written exercises. It is illustrated with many fine cuts, and brief notes at the foot of each page add greatly to its value.

Steele's Manual. (KEY TO FOURTEEN WEEKS' COURSE.)

By J. DORMAN STEELE, Ph. D. . . . . $1.00.

This is a manual of science for teachers, containing answers to the practical questions and problems in the author's scientific text-books. It also contains many valuable hints to teachers, minor tables, etc.

Wells's Science of Common Things.

By DAVID A. WELLS, A. M. . . . . $0.85.

This is a familiar explanation of the first principles of physical science for schools, families, and young students. Illustrated with numerous engravings. It is designed to furnish for the use of schools and young students an elementary text-book on the first principles of science.

Copies mailed, post-paid, on receipt of price. Full price-list sent on application.

AMERICAN BOOK COMPANY,

NEW YORK   .·.   CINCINNATI   .:·   CHICAGO.

[*70]

www.ingramcontent.com/pod-product-compliance
Lightning Source LLC
Chambersburg PA
CBHW021400210326
41599CB00011B/951